Springer Theses

Recognizing Outstanding Ph.D. Research

Aims and Scope

The series "Springer Theses" brings together a selection of the very best Ph.D. theses from around the world and across the physical sciences. Each volume has been selected for its scientific excellence by an internationally renowned institute or department. For greater accessibility to non-specialists, the published versions include an extended introduction, as well as a foreword by the student's supervisor explaining the special relevance of the work for the field. The series as a whole provides a valuable resource both for newcomers to the research fields, and for other scientists seeking detailed background information on special questions. Finally, it serves as an accredited documentation of the valuable contributions made by today's younger generation of scientists.

How to Apply

Theses prepared at internationally leading universities and institutes may be nominated for publication by supervisors with the endorsement of the head of department. Each such department is entitled to nominate one thesis per year. Enquiries and requests for a nomination form should be sent, together with the thesis, to Ms. Nathalie Berg (see link on right).

Eligibility Criteria

To qualify for consideration, theses should fulfil the following criteria:

- They must be written in good English.
- The topic should fall within the confines of Chemistry, Physics, Earth Sciences, Engineering and related interdisciplinary fields such as Materials, Nanoscience, Chemical Engineering, Complex Systems and Biophysics.
- The work reported in the thesis must represent a significant scientific advance.
- A maximum of 30% of the text may be a verbatim reproduction from the author's previous publications.
- They must have been examined and passed during the 12 months prior to nomination.
- The theses should have a clearly defined structure including an introduction accessible to new PhD students and scientists not expert in the relevant field.
- After acceptance, theses should be supplemented by a "Supervisor's Foreword" outlining the wider scientific relevance of the work

Indexed by zbMATH.

Mahmoud S. A. Alawashra

Plasma Instabilities of TeV Pair Beams Induced by Blazars

Doctoral Thesis accepted by
Institute of Physics and Astronomy, University of Potsdam,
Potsdam, Germany

Springer

Author
Dr. Mahmoud S. A. Alawashra
Astroparticle Physics Theory Group
Deutsches Elektronen-Synchrotron DESY
Zeuthen, Germany

Supervisor
Dr. Martin Pohl
Institute for Physics and Astronomy
University of Potsdam
Potsdam, Germany

Deutsches Elektronen-Synchrotron DESY
Zeuthen, Germany

ISSN 2190-5053 ISSN 2190-5061 (electronic)
Springer Theses
ISBN 978-3-031-98567-6 ISBN 978-3-031-98568-3 (eBook)
https://doi.org/10.1007/978-3-031-98568-3

This Springer imprint is published by the registered company Springer Nature Switzerland AG
The registered company address is: Gewerbestrasse 11, 6330 Cham, Switzerland

To my loving wife…

Supervisor's Foreword

One of the big questions in modern physics is the origin of magnetic field in the Universe. It is known that compression and bulk motion in galaxies and clusters of galaxies would significantly amplify a pre-existing weak magnetic field. It is this seed magnetic field whose origin is unknown. About ten years ago it was noted that in some cases the expected cascade emission of TeV-scale gamma-rays from blazars, a sub-class of active galactic nuclei, is not seen in their spectra. The underlying physical processes are fairly basic: Compton scattering, photon-photon pair production, and Lorentz invariance. Failure of model predictions based on this time-tested physics would shake the foundations of modern physics, unless a credible explanation for this failure is found. There are two proposals on the table. One involves a relatively strong magnetic field in the cosmic voids that would disperse the cascade emission. The creation of primordial magnetic field of the required strength would challenge the creativity of cosmologists. The other possible explanation argues that the TeV-scale electron-positions pairs, that are part of the cascade, drive a plasma instability in the voids that drains the beam's energy. With his thesis, Mahmoud Alawashra provides us with a very deep and carefully executed assessment of the latter option.

Earlier studies of TeV-scale pair beams and their electromagnetic cascade emission either assumed a magnetic field strong enough to deflect the pairs and hence dilute the cascade emission or probe beam-plasma instabilities in the limit of a vanishing magnetic field. In the first part of the Thesis, Mahmoud asked the relevant question: What of a weak magnetic field? It might be not strong enough to dilute the cascade emission by deflecting the pairs, but may impact the beam-plasma instabilities. Mahmoud Alawashra calculates the impact on the pair beam of a weak, structured magnetic field, and he finds that the beam would warm up for a certain range of magnetic-field strength below that needed to dilute the cascade. Then, he determines the growth rate of the fastest plasma instability for the so-modified distribution function of the pair beam, and he discovers a significant reduction in the growth rate. In a careful assessment of the nonlinear processes and their shaping of the saturation amplitude of the plasma waves, he learns that the beam-plasma interactions may no longer have the ability to suppress the gamma-ray cascade emission, leaving no viable scenario for its absence in observational data. The inevitable conclusion is that the magnetic field in the intergalactic medium may not be in this intermediate range of amplitude. This finding is very new and the result of excellent work.

In the second part of the Thesis, Mahmoud Alawashra presents a comprehensive study of the feedback on the pair beam that is imposed by the beam-plasma instability. A paper published in 2021 claimed that the primary response would be a widening of the beam akin to its warming. At the time, Mahmoud had already found that beam warming implies a reduction in the growth rate of the plasma instability. Not yet ready to accept the claim in the paper, Mahmoud set up a numerical scheme to solve both the spectral balance equation for the plasma waves and a Fokker-Planck continuity equation for the pair beam, in which the distribution function of the pair beam determines the growth rate of plasma waves, and the wave intensity shapes the Fokker-Planck coefficients. This coupling of nonlinear second-order equations requires a numerical solution of the wave dispersion relation and a careful assessment of the gridding and the numerical stability of the algorithm. Both have been executed in an exemplary fashion. It is clear that a lot of thought and attention to detail went into this part of the dissertation project. In the end Mahmoud Alawashra demonstrates that beam widening is indeed initially dominant, confirming the findings of the 2021 paper mentioned above, but that this is not generally true at later phases of evolution. There are other terms in the Fokker-Planck equation that tend to collimate the beam or provide flux to smaller kinetic energy. Taking that as an initial-value problem, Mahmoud finds the energy loss imposed by the plasma instability is at most at the per-cent level of the beam energy, even after varying the values of the most relevant parameters. Mahmoud generalizes his study by considering the continuous production of pairs and finds that a quasi-steady state can be established, for which the cascade emission is not suppressed, but the instantaneous emission is geometrically spread over about ten years. In other words, assessing the absence or presence of cascade emission in the data must be performed on the basis of the average gamma-ray flux over that timespan. The entire study is simply outstanding.

Mahmoud Alawashra's accomplishments towards the solution of a fundamental, yet mathematically very complex question elevate him far above his peers. His work represents the state-of-the-art on TeV cascades spawned by blazars. And now I invite you to read it.

Potsdam, Germany
October 2024

Dr. Martin Pohl

Abstract

Relativistic pair beams produced in the cosmic voids by TeV gamma rays from blazars are expected to produce a detectable GeV-scale cascade emission missing in the observations. The suppression of this secondary cascade implies either the deflection of the pair beam by intergalactic magnetic fields (IGMFs) or an energy loss of the beam due to the electrostatic beam-plasma instability. IGMF of femto-Gauss strength is sufficient to significantly deflect the pair beams reducing the flux of secondary cascade below the observational limits. A similar flux reduction may result in the absence of the IGMF from the beam energy loss by the instability before the inverse Compton cooling. This dissertation consists of two studies about the instability role in the evolution of blazar-induced beams.

Firstly, we investigated the effect of sub-fG level IGMF on the beam energy loss by the instability. Considering IGMF with correlation lengths smaller than a few kpc, we found that such fields increase the transverse momentum of the pair beam particles, dramatically reducing the linear growth rate of the electrostatic instability and hence the energy-loss rate of the pair beam. Our results show that the IGMF eliminates beam-plasma instability as an effective energy-loss agent at a field strength three orders of magnitude below that needed to suppress the secondary cascade emission by magnetic deflection. For intermediate-strength IGMF, we do not know a viable process to explain the observed absence of GeV-scale cascade emission and hence can be excluded.

Secondly, we probed how the beam-plasma instability feeds back on the beam, using a realistic two-dimensional beam distribution. We found that the instability broadens the beam opening angles significantly without any significant energy loss, thus confirming a recent feedback study on a simplified one-dimensional beam distribution. However, narrowing diffusion feedback of the beam particles with Lorentz factors less than 10^6 might become relevant even though initially it is negligible. Finally, when considering the continuous creation of TeV pairs, we found that the beam distribution and the wave spectrum reach a new quasi-steady state, in which the scattering of beam particles persists and the beam opening angle may increase by a factor of hundreds. This new intrinsic scattering of the cascade can result in time delays of around ten years, thus potentially mimicking the IGMF deflection.

Understanding the implications on the GeV cascade emission requires accounting for inverse Compton cooling and simulating the beam-plasma system at different points in the IGM.

List of Publications

Peer-Reviewed Publications

- Mahmoud Alawashra and Martin Pohl. Suppression of the TeV Pair-beam–Plasma Instability by a Tangled Weak Intergalactic Magnetic Field. **The Astrophysical Journal**, 929(1): 67, 2022. https://dx.doi.org/10.3847/1538-4357/ac5a4b
- Mahmoud Alawashra and Martin Pohl. Nonlinear Feedback of the Electrostatic Instability on the Blazar-induced Pair Beam and GeV Cascade. **The Astrophysical Journal**, 964(1): 82, 2024. https://dx.doi.org/10.3847/1538-4357/ad24ea

Conference Proceedings

- Mahmoud Alawashra and Martin Pohl. Suppression of the TeV pair-beam plasma instability by a weak intergalactic magnetic field. **PoS** 395—37th International Cosmic Ray Conference (ICRC2021). https://doi.org/10.22323/1.395.0990
- Mahmoud Alawashra and Martin Pohl. Nonlinear feedback of the electrostatic instability on the blazar-induced pair beam and GeV cascade. **PoS** 444—38th International Cosmic Ray Conference (ICRC2023). https://doi.org/10.22323/1.444.1456

Acknowledgements

I am profoundly grateful to my advisor, Prof. Martin Pohl, for his unwavering support, guidance, and mentorship throughout this incredible research journey. His expertise, constructive criticism, and encouragement have been invaluable in shaping the direction and quality of this work. I would like to thank my co-supervisor, Dr. Kfir Blum, and my mentor, Dr. Andrew Taylor, for the interesting discussions during my Ph.D. committee meetings and for their comments on my papers. I extend my appreciation to the reviewers, Prof. Christoph Pfrommer and Prof. Antoine Bret, the chairman, Prof. Stephan Geier, and the committee members, Prof. Achim Feldmeier and Prof. Tim Dietrich.

Special appreciation is acknowledged to Dr. Michelle Tsirou and Dr. Jan Benácek for reviewing the second and third chapters of the thesis and to Karol Fulat for the interesting discussions across our shared office. I am also grateful to all my peers and colleagues at the University of Potsdam and DESY. I am very thankful to Ines Tietgen and Silvia Alessandria for their genuine support on matters of administration. I would like to acknowledge the financial support provided by the International Helmholtz-Weizmann Research School for Multimessenger Astronomy, largely funded through the Initiative and Networking Fund of the Helmholtz Association.

I cannot adequately convey my gratitude to my wife, Dana. I sincerely appreciate her steadfast support throughout the years, both during challenging moments and joyful occasions. Her presence has consistently been my greatest source of comfort. Last but not least, I extend my sincere gratitude to my family for their encouragement. Their unwavering belief in my capabilities and their patience during the challenging stages of this journey has been the foundation of my resilience and determination.

Contents

List of Figures

List of Tables

Chapter 1
Introduction

Gamma-ray astronomy is a recent cutting-edge astrophysics field that unveils the universe's most extreme and enigmatic phenomena. Employing the power of high-energy gamma rays that emerge from the most violent cosmic processes in the universe this discipline provides a unique lens through which we can explore supernovae, pulsars, gamma-ray bursts, and the active galactic nuclei (AGNi) [1]. AGNi stand as the brightest persistent emitters of electromagnetic radiation in the universe [2]. Consequently, they serve as valuable tools for discovering distant objects and their evolution over cosmic epochs and provide constraints on our models of the universe's evolution.

Blazars are AGNi with their relativistic jet pointing toward Earth. Observations of the Fermi-LAT telescope and the imaging atmospheric Cerenkov telescopes (such as VERITAS, MAGIC, and H.E.S.S.) show a bright GeV-TeV gamma-ray emission of several blazars. During their propagation through the intergalactic medium (IGM), those very high energy gamma-rays interact with the extragalactic background light (EBL) photons, producing a focused beam of electron-positron pairs with a wide range of Lorentz factors between 10^3 and 10^8. These pairs are anticipated to dissipate their energies via inverse Compton scattering on the cosmic microwave background (CMB) giving a secondary GeV-scale gamma-ray cascade that is detectable by Fermi-LAT telescope [3, 4].

Although primary blazar gamma-rays with energies of a few TeV would initiate an electromagnetic cascade in the GeV energy range, such emissions seem to be absent from the gamma-ray spectra of certain blazars [5] and possibly the isotropic gamma-ray background [6]. One possible explanation for the absence of the GeV cascade emission from the gamma-ray spectra of blazars is the TeV pairs deflection by the intergalactic magnetic fields (IGMFs) [7–13]. This deflection results in an extended emission or/and a time delay of the cascade emission. In this case, the observed blazar spectra are used to put lower limits on the strength of the IGMFs.

The field strength required to suppress the cascade emission due to the time delay is around $B_{\mathrm{IGM}} > 10^{-15}$G for IGMF with a correlation length similar to or larger than the energy loss length of the beam, $\lambda_{\mathrm{B}} \gtrsim 10$ kpc, and stronger than that for a

M. S. A. Alawashra, *Plasma Instabilities of TeV Pair Beams Induced by Blazars*, Springer Theses, https://doi.org/10.1007/978-3-031-98568-3_1

small correlation length, for which the beam sees a fluctuating magnetic field and the deflection becomes diffusive [14]. If the magnetic field is strong enough to deflect by a radian or more, then the cascade emission from AGNi with oblique jets (jets more than 30° off of our line of sight) should become visible [15, 16], but corresponding emission has not been found [14, 17, 18].

The only alternative solution for the missing GeV cascade emission within the standard model physics is the beam energy loss by the collective beam-plasma instabilities before the inverse Compton cooling on the CMB. However, whether the non-linear evolution of those instabilities is efficient in taking significant fraction of the beam energy is still debatable in the literature [19–33]. If the nonlinear evolution of the instability is efficient in draining the energy from the beam particles then the IGM could be subject to significant heating. Cosmological simulations including this heating process can successfully reproduce the observed IGM temperature and the effective optical depth as a function of redshift as well as several other observations [34].

Vafin et al. [30] calculated the linear growth rate of the electrostatic instability using a realistic pair distribution generated by the annihilation of high-energy gamma rays with the extragalactic background light. Their results demonstrated that the finite angular spread of the beam plays a decisive role in shaping the unstable electrostatic modes. They concluded that the balance of instability growth with the modulation instability damping determines the saturation level, yielding an energy loss rate of the beam due to the instability that is faster than that due to the comptonization on the CMB photons.

The beam-plasma electrostatic instability operates best in the absence of a magnetic field. Noting that magnetic deflection needs more than a femto-Gauss field amplitude, here in the first study related to this thesis [35] and presented in Chap. 4, we address the effect on the electrostatic instability that would be imposed by much weaker intergalactic magnetic fields with a correlation length much smaller than the beam energy loss length. In particular, we investigate whether the beam-plasma instability still is the dominant energy-loss process and how strongly the cascade emission is suppressed.

Apart from the non-linear wave interactions, there is also the non-linear feedback of the waves on the beam. Previous studies of the blazar-induced pair beam instabilities didn't consider this feedback of the instability. This feedback has been studied for the first time in the context of blazar-induced pair beam electrostatic instability by Perry and Lyubarsky [36]. Their findings imply that the back reaction of the electrostatic unstable waves on the pair beam widens the beam opening angles by around one order of magnitude without any significant energy loss.

In the second work related to this thesis [37] and presented in Chap. 5, we use a two-dimensional realistic beam distribution to explore the influence of the instability feedback on the beam. This is unlike the simplified one-dimensional beam distribution used in Perry and Lyubarsky [36]. Specifically, we use the beam profile at a distance of 50 Mpc from the blazar found in Vafin et al. [30]. This treatment enables us to look at the feedback influence on the pairs that have the relevant Lorentz factors for an inverse Compton scattering cascade in the GeV band.

The instability feedback is described by a Fokker-Planck diffusion both in momentum and angular space. This treatment was simplified in the analysis by Perry and Lyubarsky [36], by evaluating only the initially dominant angular widening diffusion and neglecting the other subdominant feedback involving the momentum diffusion and angular narrowing diffusion. Here, we check rigorously this assumption by using the 2D spectrum of the expanded beam under the dominant feedback to analyse the possible impact of the momentum diffusion on the beam energy and whether the beam narrowing diffusion is still negligible.

The beam-plasma instability significantly outpaces other factors that could change the blazar-induced pair beam profile, such as inverse Compton cooling and pair production. Whereas previous works have predominantly focused on assessing the instability's impact on a stationary beam profile, we incorporate in Chap. 4 the continuous production of TeV pairs into the transport equation of the beam, in addition to the instability feedback on the beam.

We start in Chap. 2 with an overview of the field of gamma-ray astronomy, where we will introduce the main persistent extragalactic sources of the gamma-ray mainly AGNi and their subclass blazars. We also follow the gamma-ray journey across the cosmic voids from those sources to Earth. In Chap. 3, we take a close look at the foundations and the relevant physics of the beam-plasma instabilities including their non-linear interactions. The following Chap. 4 focus on our first work where we investigated the impact of weak IGMF on the growth of the beam-plasma instability, we derived the magnetic fields limit at which the instability is suppressed by the tangled IGMF. Later, in Chap. 5, we present our second study and its methodology where we simulated the Fokker-Planck diffusion feedback of the instability on the beam. Ultimately we conclude and give an outlook of further possible investigations in Chap. 6.

References

1. P. V. Ramana Murthy and A. W. Wolfendale. *Gamma-ray Astronomy*. Cambridge Astrophysics. Cambridge University Press, 2 edition, 1993
2. P. Padovani, D. M. Alexander, R. J. Assef, B. De Marco, P. Giommi, R. C. Hickox, G. T. Richards, V. Smolčić, E. Hatziminaoglou, V. Mainieri, and M. Salvato. Active galactic nuclei: what's in a name? *Astron Astrophys Rev*, 25(2), 2017. https://doi.org/10.1007/s00159-017-0102-9
3. Gould Robert J, Schréder Gérard P (March 1967) Opacity of the Universe to High-Energy Photons. Physical Review 155(5):1408–1411. https://doi.org/10.1103/PhysRev.155.1408
4. Blumenthal George R, Gould Robert J (January 1970) Bremsstrahlung, Synchrotron Radiation, and Compton Scattering of High-Energy Electrons Traversing Dilute Gases. Reviews of Modern Physics 42(2):237–271. https://doi.org/10.1103/RevModPhys.42.237
5. Neronov A, Semikoz DV (December 2009) Sensitivity of γ-ray telescopes for detection of magnetic fields in the intergalactic medium. Physical Review D 80(12):123012. https://doi.org/10.1103/PhysRevD.80.123012
6. Carlos Blanco, Oindrila Ghosh, Sunniva Jacobsen, and Tim Linden. Where are the Cascades from Blazar Jets? An Emerging Tension in the γ-ray sky. *arXiv e-prints*, art. arXiv:2303.01524, March 2023. https://doi.org/10.48550/arXiv.2303.01524

7. Elyiv A, Neronov A, Semikoz DV (Jul2009) Gamma-ray induced cascades and magnetic fields in the intergalactic medium. Phys. Rev. D 80:023010. https://link.aps.org/doi/10.1103/PhysRevD.80.023010
8. Neronov A, Semikoz DV (Dec2009) Sensitivity of γ-ray telescopes for detection of magnetic fields in the intergalactic medium. Phys. Rev. D 80:123012. https://link.aps.org/doi/10.1103/PhysRevD.80.123012
9. Neronov Andrii, Vovk Ievgen (April 2010) Evidence for Strong Extragalactic Magnetic Fields from Fermi Observations of TeV Blazars. Science 328(5974):73. https://doi.org/10.1126/science.1184192
10. Taylor AM, Vovk I, Neronov A (May2011) Extragalactic magnetic fields constraints from simultaneous GeV-TeV observations of blazars. A&A 529:A144. https://doi.org/10.1051/0004-6361/201116441
11. Takahashi Keitaro, Mori Masaki, Ichiki Kiyotomo, Inoue Susumu (Dec2011) LOWER BOUNDS ON INTERGALACTIC MAGNETIC FIELDS FROM SIMULTANEOUSLY OBSERVED GeV-TeV LIGHT CURVES OF THE BLAZAR mrk 501. The Astrophysical Journal 744(1):L7. https://doi.org/10.1088/2041-8205/744/1/l7
12. Ie. Vovk, A. M. Taylor, D. Semikoz, and A. Neronov. FERMI /LAT OBSERVATIONS OF 1es 0229+200: IMPLICATIONS FOR EXTRAGALACTIC MAGNETIC FIELDS AND BACKGROUND LIGHT. *The Astrophysical Journal*, 747(1):L14, feb 2012. https://doi.org/10.1088/2041-8205/747/1/l14
13. Ruth Durrer and Andrii Neronov. Cosmological magnetic fields: their generation, evolution and observation. *The Astronomy and Astrophysics Review*, 21(1), Jun 2013. ISSN 1432-0754. https://doi.org/10.1007/s00159-013-0062-7
14. M. Ackermann, M. Ajello, L. Baldini, J. Ballet, G. Barbiellini, D. Bastieri, R. Bellazzini, E. Bissaldi, R. D. Blandford, E. D. Bloom, and et al. The search for spatial extension in high-latitude sources detected by the fermi large area telescope. *The Astrophysical Journal Supplement Series*, 237(2):32, Aug 2018. ISSN 1538-4365. https://doi.org/10.3847/1538-4365/aacdf7
15. Broderick Avery E, Tiede Paul, Shalaby Mohamad, Pfrommer Christoph, Puchwein Ewald, Chang Philip, Lamberts Astrid (2016) Bow Ties in the Sky I: The Angular Structure of Inverse Compton Gamma-ray Halos in the Fermi Sky. Astrophys. J. 832(2):109. https://doi.org/10.3847/0004-637X/832/2/109
16. Tiede Paul, Broderick Avery E, Shalaby Mohamad, Pfrommer Christoph, Puchwein Ewald, Chang Philip, Lamberts Astrid (April 2020) Constraints on the Intergalactic Magnetic Field from Bow Ties in the Gamma-Ray Sky. ApJ 892(2):123. https://doi.org/10.3847/1538-4357/ab737e
17. Paul Tiede, Avery E. Broderick, Mohamad Shalaby, Christoph Pfrommer, Ewald Puchwein, Philip Chang, and Astrid Lamberts. Bow Ties in the Sky. II. Searching for Gamma-Ray Halos in the Fermi Sky Using Anisotropy. *Astrophys. J.*, 850(2):157, 2017. https://doi.org/10.3847/1538-4357/aa9375
18. Broderick Avery E, Tiede Paul, Chang Philip, Lamberts Astrid, Pfrommer Christoph, Puchwein Ewald, Shalaby Mohamad, Werhahn Maria (Nov2018) Missing gamma-ray halos and the need for new physics in the gamma-ray sky. The Astrophysical Journal 868(2):87. https://doi.org/10.3847/1538-4357/aae5f2
19. Avery E. Broderick, Philip Chang, and Christoph Pfrommer. THE COSMOLOGICAL IMPACT OF LUMINOUS TeV BLAZARS. i. IMPLICATIONS OF PLASMA INSTABILITIES FOR THE INTERGALACTIC MAGNETIC FIELD AND EXTRAGALACTIC GAMMA-RAY BACKGROUND. *The Astrophysical Journal*, 752(1):22, may 2012. https://doi.org/10.1088/0004-637x/752/1/22
20. Schlickeiser R, Ibscher D, Supsar M (Oct2012) PLASMA EFFECTS ON FAST PAIR BEAMS IN COSMIC VOIDS. The Astrophysical Journal 758(2):102. https://doi.org/10.1088/0004-637x/758/2/102
21. Miniati Francesco, Elyiv Andrii (May2013) RELAXATION OF BLAZAR-INDUCED PAIR BEAMS IN COSMIC VOIDS. The Astrophysical Journal 770(1):54. https://doi.org/10.1088/0004-637x/770/1/54

22. R. Schlickeiser, S. Krakau, and M. Supsar. PLASMA EFFECTS ON FAST PAIR BEAMS. II. REACTIVE VERSUS KINETIC INSTABILITY OF PARALLEL ELECTROSTATIC WAVES. *The Astrophysical Journal*, 777(1):49, oct 2013. https://doi.org/10.1088/0004-637x/777/1/49
23. Broderick Avery E, Pfrommer Christoph, Puchwein Ewald, Chang Philip (August 2014) Implications of Plasma Beam Instabilities for the Statistics of the Fermi Hard Gamma-Ray Blazars and the Origin of the Extragalactic Gamma-Ray Background. ApJ 790(2):137. https://doi.org/10.1088/0004-637X/790/2/137
24. Sironi Lorenzo, Giannios Dimitrios (May2014) RELATIVISTIC PAIR BEAMS FROM TeV BLAZARS: A SOURCE OF REPROCESSED GeV EMISSION RATHER THAN INTERGALACTIC HEATING. The Astrophysical Journal 787(1):49. https://doi.org/10.1088/0004-637x/787/1/49
25. Chang Philip, Broderick Avery E, Pfrommer Christoph, Puchwein Ewald, Lamberts Astrid, Shalaby Mohamad (Dec2014) THE EFFECT OF NONLINEAR LANDAU DAMPING ON ULTRARELATIVISTIC BEAM PLASMA INSTABILITIES. The Astrophysical Journal 797(2):110. https://doi.org/10.1088/0004-637x/797/2/110
26. Markus Supsar and Reinhard Schlickeiser. PLASMA EFFECTS ON FAST PAIR BEAMS. III. OBLIQUE ELECTROSTATIC GROWTH RATES FOR PERPENDICULAR MAXWELLIAN PAIR BEAMS. *The Astrophysical Journal*, 783(2):96, feb 2014. https://doi.org/10.1088/0004-637x/783/2/96
27. Chang Philip, Broderick Avery E, Pfrommer Christoph, Puchwein Ewald, Lamberts Astrid, Shalaby Mohamad, Vasil Geoffrey (Dec2016) The linear instability of dilute ultrarelativistic e±pair beams. The Astrophysical Journal 833(1):118. https://doi.org/10.3847/1538-4357/833/1/118
28. Kempf A, Kilian P, Spanier F (January 2016) Energy loss in intergalactic pair beams: Particle-in-cell simulation. Astronomy & Astrophysics 585:A132. https://doi.org/10.1051/0004-6361/201527521
29. Rafighi I, Vafin S, Pohl M, Niemiec J (November 2017) Plasma effects on relativistic pair beams from TeV blazars: PIC simulations and analytical predictions. Astronomy & Astrophysics 607:A112. https://doi.org/10.1051/0004-6361/201731127
30. S. Vafin, I. Rafighi, M. Pohl, and J. Niemiec. The electrostatic instability for realistic pair distributions in blazar/ebl cascades. *The Astrophysical Journal*, 857(1):43, Apr 2018. ISSN 1538-4357. https://doi.org/10.3847/1538-4357/aab552
31. Vafin S, Deka PJ, Pohl M, Bohdan A (Feb2019) Revisit of nonlinear landau damping for electrostatic instability driven by blazar-induced pair beams. The Astrophysical Journal 873(1):10. https://doi.org/10.3847/1538-4357/ab017b
32. Rafael Alves Batista, Andrey Saveliev, and Elisabete M. de Gouveia Dal Pino. The Impact of Plasma Instabilities on the Spectra of TeV Blazars. *Mon. Not. Roy. Astron. Soc.*, 489(3):3836–3849, 2019. https://doi.org/10.1093/mnras/stz2389
33. Shalaby Mohamad, Broderick Avery E, Chang Philip, Pfrommer Christoph, Puchwein Ewald, Lamberts Astrid (2020) The growth of the longitudinal beam–plasma instability in the presence of an inhomogeneous background. Journal of Plasma Physics 86(2):535860201. https://doi.org/10.1017/S0022377820000215
34. Puchwein Ewald, Pfrommer Christoph, Springel Volker, Broderick Avery E, Chang Philip (2012) The Lyman-alpha forest in a blazar-heated Universe. Mon. Not. Roy. Astron. Soc. 423:149. https://doi.org/10.1111/j.1365-2966.2012.20738.x
35. Alawashra Mahmoud, Pohl Martin (Apr2022) Suppression of the TeV Pair-beam–Plasma Instability by a Tangled Weak Intergalactic Magnetic Field. The Astrophysical Journal 929(1):67. https://doi.org/10.3847/1538-4357/ac5a4b
36. Perry Roy, Lyubarsky Yuri (May2021) The role of resonant plasma instabilities in the evolution of blazar-induced pair beams. MNRAS 503(2):2215–2228. https://doi.org/10.1093/mnras/stab324
37. Alawashra Mahmoud, Pohl Martin (Mar2024) Nonlinear feedback of the electrostatic instability on the blazar-induced pair beam and gev cascade. The Astrophysical Journal 964(1):82. https://doi.org/10.3847/1538-4357/ad24ea

Chapter 2
Observational Context

In this chapter, we present several concepts that are relevant to the scope of this thesis: studying the beam-plasma instability of the blazar-induced TeV pair beams. We discuss the extragalactic TeV gamma-ray sky in Sect. 2.1. In Sect. 2.2, we introduce the active galactic nuclei (AGNi) and Blazars which are astrophysical sources producing TeV-gamma rays propagating in the intergalactic medium (IGM). Later in Sect. 2.3, we introduce the IGM environment where those very-high-energy (VHE) gamma rays propagate in before reaching us. Finally, we follow the TeV gamma rays journey through the comics voids of the IGM in Sect. 2.4.

2.1 The Extragalactic Gamma-Ray Sky

The thermal radiation in the Universe is emitted by several astrophysical objects like stars, dust and gas. This radiation lies mainly in the optical and infrared wavelengths. However, along with this thermal radiation, we also observe non-thermal radiation at several wavelengths like radio, X-ray and gamma-ray. This non-thermal emission is possible by the acceleration of charged particles through extreme processes in astrophysical environments. Those high-energy charged particles radiate by several emission mechanisms like the synchrotron radiation when spiralling magnetic fields or the inverse Compton scattering on low-energy target photons [11].

The field of gamma-ray astronomy is a relatively young disciplinary field compared to thermal astronomy. Unlike the detection of thermal light, the detection of gamma rays presents different challenges and requires different techniques. Gamma rays with very high energies (100 GeV–100 TeV) are observed by the ground-based Imaging Atmospheric Cherenkov Telescopes IACTs (i.e H.E.S.S., MAGIC, VERITAS). The basic idea behind this detection technique is to detect the flash of Cherenkov radiation induced by the cascade of charged particles resulting from the interaction of the gamma-ray with the atmosphere. This flash illuminates an area of

M. S. A. Alawashra, *Plasma Instabilities of TeV Pair Beams Induced by Blazars*,
Springer Theses, https://doi.org/10.1007/978-3-031-98568-3_2

hundreds of square meters on the ground, this requires distances of hundred meters between the telescopes to ensure the coverage of the cascade [12].

Other gamma-ray telescopes include space-based telescopes. Where a tracker and a calorimeter are used to detect the shower generated by the gamma-ray. Those particle-physics instruments are carried on the board of a satellite orbiting Earth like the Fermi gamma-ray space telescope. This technique is mostly effective at the detection of gamma rays with energies of 50 MeV up to a few hundred GeV. The small effective area of this technique compared to the ground-based telescopes is the resonance for the limitation of a few hundred GeV on the maximum detected energy.

The field of Extragalactic gamma-ray astronomy started during the 1990s, in Fig. 2.1 we see the number of observed extragalactic gamma-ray sources with time. EGRET (Energetic Gamma Ray Experiment Telescope) detector was one of the first space-based gamma-ray telescopes during the 1990s. The detector on board the Compton Gamma-Ray Observatory revealed almost a hundred bight extragalactic sources with gamma-ray energies above 100 MeV band like Flat Spectrum Radio Quasars. This telescope also allowed for the firm creation of the entire gamma-ray sky map, where the comparison with multi-wavelength observations revealed that the majority of those sources are Active Galactic Nuclei (AGNi). Those objects are the host galaxies of a supermassive black hole with an active accreting.

During the early 2000s, the Whipple Observatory, a ground-based gamma-ray telescope, detected seven blazars (AGN with their jet closely aligned with the line of sight) and a single radio galaxy, M87. Later after 2005, a new generation of ground-based gamma-ray telescopes emerged, VERITAS in Arizona, USA, MAGIC in La Palma, Canary Islands and H.E.S.S. in Khomas Highlands, Namibia. With enhanced imaging capabilities along with improved background rejection, they discovered a population of TeV extragalactic sources called BL Lac objects (BLL). BL Lac is a type of blazar with Spectral Energy Distributions that peaked at very high energies. Those telescopes also revealed TeV emissions from galaxies surrounding the Local Group, where the Milky Way is located, like M82, NGC 253 and Centaurus A.

In 2008, the Fermi Large Area Telescope (Fermi-LAT) was launched with an effective area of around a square meter and a field of view larger by about five times than EGRET [13]. This instrument has given rise to an enormous increase in the number of observed extragalactic gamma-ray sources with energies above a hundred MeV as we can see in Fig. 2.1. Thanks to Fermi-LAT angular resolution of around a degree for energies above GeV, the comparison with observations of x-ray, optical and radio was possible. Having multi-wavelength observations has provided us with extraordinary insights into the violent process happening in the environments of extragalactic gamma-ray sources. In the next section, we will take a look at one of the prominent gamma-ray astrophysical objects called blazars.

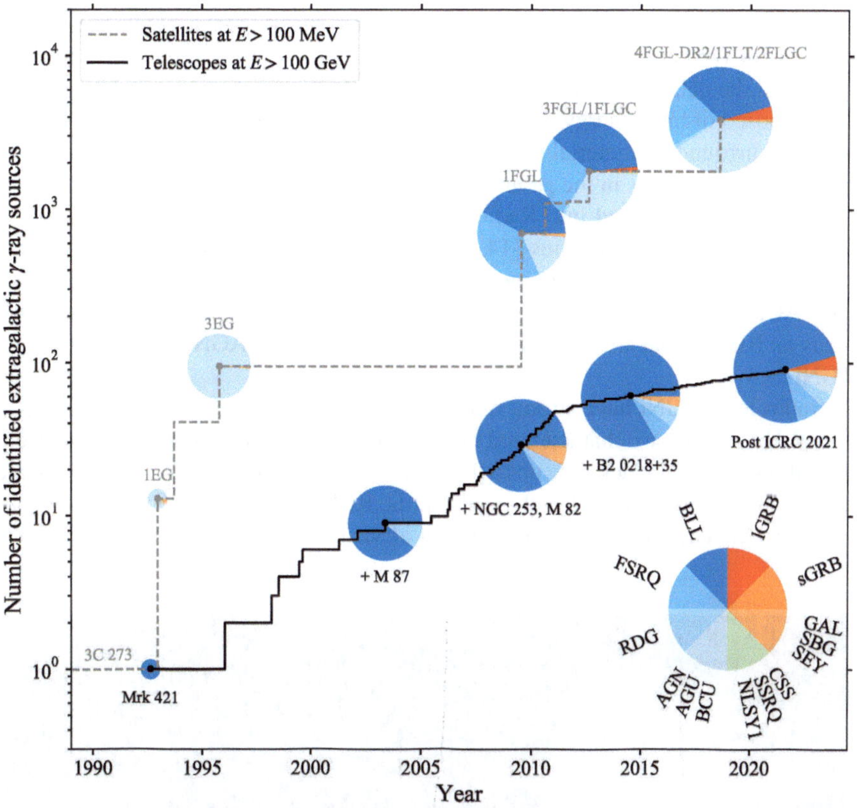

Fig. 2.1 The number of detected extragalactic gamma-ray sources with time. Space-based telescopes (like EGRET and Fermi-LAT) detections are shown by the dashed grey line. Those detections are reported from the following catalogues: 1EG [1], 3EG [2], 1FGL [3], 3FGL [4], 1FLGC [5], 4FGL-DR2 [6], 1FLT [7] and 2FLGC [8]. The solid black line shows the detections by ground-based gamma-ray telescopes as reported by TeVCat [9] during the 2021 International cosmic-ray Conference (ICRC 2021). The categories of the sources are displayed in the pie chart at the lower right of the Figure. Blue: BL Lac (BLL), flat-spectrum radio quasar (FSRQ), radio galaxy (RDG) and active galactic nuclei of either radio galaxy of blazar-type with uncertain classification (AGN, AGU, BCU). Green: Narrow-line Seyfert 1 (NLSY1), steep-spectrum radio quasars (SSRQ) and compact steep spectrum sources (CSS). Orange: regular star-forming galaxies along with starburst galaxies and non-jetted Seyfert-type AGN (GAL, SBG, SEY), gamma-ray bursts of short (sGRB) and long (lGRB) duration. Figure was taken from Biteau and Meyer [10]

2.2 AGNi and Blazars

Blazars are a class of active galactic nuclei (AGNi) with a relativistic jet pointing along the line of sight of the Earth. They are characterized by intense and persistent radiation across the electromagnetic spectrum, ranging from radio waves to high-energy TeV gamma-rays.

AGNi are galaxies with an actively accreting supermassive black hole at the centre with typical masses of $10^7 - 10^{10}$ solar masses. They represent a few per cent of all the galaxies in the observed universe. AGNi are characterized by a bright jet composed of high-energy relativistic particles that emit radiation as they interact with the surrounding material and magnetic fields. Over half of the electromagnetic luminosity of AGNi is in the gamma-rays band ($E > 100$ MeV).

AGNi are categorized into Radio-quiet AGNi and Radio-loud AGNi based on their radio emission properties [14, 15]. Radio-quiet AGNi have relatively weak radio emissions, with radio luminosities $L_{5\text{GHz}} < 10^{25}$ WHz^{-1}. Examples of radio-quiet AGNi include Seyfert galaxies and Radio-quiet quasars/QSOs. In this thesis, we focus on Blazars that are classified within the other type of AGNi called Radio-loud AGNi.

Radio-loud AGNi have strong radio emission, with radio luminosities $L_{5\text{GHz}} > 10^{26}$ WHz^{-1}. They are characterized by prominent relativistic jets extending out from the galaxy's central black hole up to the intergalactic medium. In Fig. 2.2, we show a multi-wavelength picture of such a galaxy, the Radio galaxy 3C 348 (also known as Hercules A). The picture is a combination of visible wavelengths by Hubble Space

Fig. 2.2 Multi-wavelength picture of the Radio galaxy 3C 348 (also known as Hercules A). Hubble Space Telescope's Wide Field Camera 3 reviles the yellowish elliptical galaxy at the centre in visible wavelengths of light, while the Very Large Array (VLA) radio telescope reviles the jetted outflow of the galaxy in radio wavelengths shown in pink. [Credits: NASA, ESA, S. Baum and C. O'Dea (RIT), R. Perley and W. Cotton (NRAO/AUI/NSF), and the Hubble Heritage Team (STScI/AURA), hubblesite.org]

Telescope's Wide Field Camera 3 and radio wavelengths by the Very Large Array (VLA) radio telescope in New Mexico. In the radio component, we see clearly the very high-energy plasma beams of the jet of the supermassive black hole at the centre of the galaxy. This jet extends for a distance of around one million light-years in the IGM.

AGNi are also classified based on the orientation of the accretion disk and the jet relative to the observer. This is the so-called unification model of the AGNi that was introduced by Urry and Padovani [16] for the radio-loud ones and by Antonucci [17] for radio-quiet sources. In Fig. 2.3, this unification is illustrated [18]. This Figure also shows that the presence or absence of radio jets is the difference between radio-loud and radio-quiet AGNi. Finally, it is important to highlight that there are more classifications in the literature, for example, based on the evolution of the host galaxy of the AGN [19, 20].

When the jet of the radio-loud galaxy is closely aligned with our line of sight (Blazars), we observe an enhancement in the energy of the photon compared to the emitted photon in the jet frame. This is due to the relativistic beaming of the photons in the jet outflow with an opening angle of $\theta \sim \Gamma^{-1}$, where Γ is the bulk Lorentz factor of the jet plasma outflow. The result is an increase in the observed luminosity

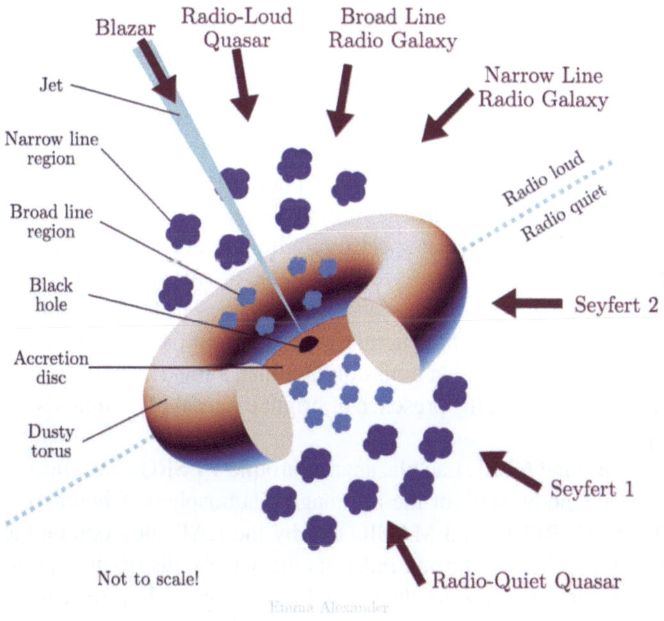

Fig. 2.3 Representation of the unified AGN model. The classification of the AGN depends on the viewing angle we see the object, and whether or not there is a significant jet emission. In particular, we see that blazars are a subclass of Radio loud AGN where the jet points toward us [Credit: Emma Alexander, emmaalexander.github.io]

of the jet by a factor of δ^4, where δ is the relativistic Doppler Factor of the jet that is of the order 10 [21]. This Lorentz boosting makes it possible to detect the weak jet emission of blazars.

The spectral energy distribution (SED) of blazars stretches over the entire electromagnetic spectrum from radio up to very high-energy gamma-ray. This SED has two spectral humps, the first peaks between the optical and X-ray bands, whereas the higher energy one peaks in the gamma-ray band. The low-energy bump is usually assigned to the beamed synchrotron emission from relativistic electrons or pairs in the jet, where it is possible to interpret the high-energy one with leptonic or hadronic models. In the leptonic approach, the emission is explained as the radiation produced via inverse Compton scattering of the same electrons or pairs that have produced the synchrotron radiation on target lower energy photons. The target photons for this scattering can be the synchrotron radiation photons or external photons from the surrounding [22–25].

On the other hand, along with the leptonic component, a hadronic one might contribute to the blazar emission. In the case of purely hadronic models [26], all the gamma-ray emission can be traced to the hadronic interactions and synchrotron radiation either from the secondary leptons created in hadronic processes or from protons. However, extreme jet powers are required normally in those scenarios [27].

In the case of lepto-hadronic models, the low-energy peak of the blazar SED is explained by the synchrotron emission of electrons as in the leptonic models. Whereas the high-energy emission has a hadronic component along with a leptonic one. In the case of strong magnetic fields 1-100 Gauss, synchrotron radion from protons with energies up to 10^{19} eV dominates the high-energy peak [28]. In the case of smaller magnetic fields and large particle and photon field densities, synchrotron emission from the p-γ induced cascades (i.e. photomeson processes) could dominate the high-energy emission [28, 29].

Blazars are classified into flat-spectrum radio quasars (FSRQ) and BL Lacertae objects (BLL). BL Lac blazars have very weak or missing emission lines while the FSRQs are characterized by broad emission lines in their optical spectra. BL Lacs also in general have higher energies of peak luminosity of the emitted gamma rays [30]. The inverse Compton emission dominates the SED of the FSRQs due to the high accretion rate and the presence of bright external photon fields compared to the BLLs [24].

Up to now, around 67 BL Lac blazars and around 9 FSRQs have been detected in the gamma-ray band by state-of-the-art imaging atmospheric Cherenkov telescopes such as HESS, VERITAS and MAGIC and by the LAT telescope on the board of Fermi satellite [9]. Those Blazars redshifts are widely distributed, up to redshifts of 1 for the blazars observed by the ground telescopes and up to redshifts of 4 for the ones observed by the space-based telescopes. The fact that we observe gamma rays from such a distance gives us a unique way to study the interactions of gamma rays during their propagation in cosmic voids. We will present those interactions in Sect. 2.4 and follow the gamma-ray journey in cosmic voids, however, before this, we introduce the cosmic voids and their radiation content in the following section.

2.3 Intergalactic Medium

Most of the universe consists of voids between galaxies called the "intergalactic medium" (IGM). This medium has a density much lower than the lowest density of vacuum we have made in the lab which is about 1000 atoms per cubic centimetre achieved at CERN. Typical densities of the electrons in the intergalactic medium are of the order ($n_e \approx 10^{-7}\ \mathrm{cm}^{-3}$) where z is the redshift [31], while the IGM temperature is of the order 10^4 Kelvin [32]. We will be adapting those values of the IGM density and temperature throughout our work.

The IGM contains half of the dark matter in the universe, whereas the present fraction of baryons that are intergalactic is probably much higher [33]. Baryonic matter within the IGM has been leaked out of galaxies due to the interactions between galaxies, star formation processes, and the influence of supermassive black holes. It is also thought that the star and supermassive black hole formation are the primary sources of the IGM heating, metal enrichment and ionization [33].

Cosmic microwave background (CMB) observations indicate that cosmic gas recombined and turned neutral at around the Universe's 400,000th year ($z \approx 1100$), marking the end of a recombination era extending up to those redshifts. After that, it is anticipated that the intergalactic medium (IGM) will maintain its neutrality until radiation sources capable of reionization emerge [33]. As the Universe expanded, the cosmic gas adiabatically cooled, leading to the emergence of the intergalactic medium (IGM), whose structure primarily arises from the gravitational influence on primordial matter fluctuations.

The inception of star formation in the Universe likely began within the exceptionally rare peak of the cosmological density field at a redshift of approximately $z \sim 70$ [34]. These early stars are thought to have formed within halos with masses spanning from 10^5 to 10^7 solar masses. Over time, the universe witnessed a progressive increase in star formation, ultimately reaching a saturation level in the star formation rate of the Population III stars in minihalos at redshifts of around $z \sim 20 - 30$ [35]. Population III stars are considered to be the main source of radiative feedback in the universe at redshifts ($z > 22$) and the first luminous objects to form during the dark ages of the universe when hydrogen remained in a neutral state [36].

As the universe was just a few hundred million years old, the first galaxies formed marking the end of the cosmic dark ages. The majority of them could already have contained low-mass, Population II stars and possibly stellar clusters [36–38]. The intergalactic medium, which just started to emerge, was directly affected by radiative backgrounds associated with the first galaxies. In the beginning, the IGM gas was heated by roughly 10 eV and soft X-ray background photons such that the 21 cm line of atomic hydrogen could potentially be detected [39, 40]. Following this, photons with energies greater than 13.6 eV photoionized approximately all the intergalactic hydrogen, a process known as cosmological "reionization", which in turn raised the temperature of the IGM to tens of thousands of degrees Kelvin.

At redshifts $2 < z < 5$, the IGM can be probed by the hydrogen Lyα absorption line that has been redshifted to the point that it can be observed by the ground-based optical telescopes [41–43]. And because the light encounters several IGM gas clouds at different redshifts, multiple Lyα absorption lines are usually formed in the observed quasar spectra in what is known as the Lyα forest spectral region. Hundreds of well-resolved such Lyα forest spectral regions have been observed in quasars spectra by the 10-meter telescopes and a hundred thousand with medium resolution by the 2.5-meter Sloan telescope [44, 45].

This wealth of absorption line data for the IGM gas at those redshifts ($2 < z < 5$) is crucial for our understanding of IGM. It has been employed for example to determine the thermal history of the IGM [46–51] and constrain the history of intergalactic enrichment [52, 53]. However, it's difficult to observe the IGM gas at redshifts lower than two since the expansion of the universe has diluted most cosmic gas by those redshifts.

Redshift surveys of galaxies revealed the cosmic web's large-scale structures where clusters, that extend over several megaparsecs in radius, are connected by extensive filaments spanning numerous tens of megaparsecs. Those filaments envelop underdense voids, with radii measuring tens of megaparsecs. These large-scale structures cosmic web have evolved through gravitational instabilities from the initial density fluctuations of the early Universe [55]. In Fig. 2.4a, we see this large-scale structure of the universe in the Sloan Digital Sky Survey (SDSS) galaxy map [56].

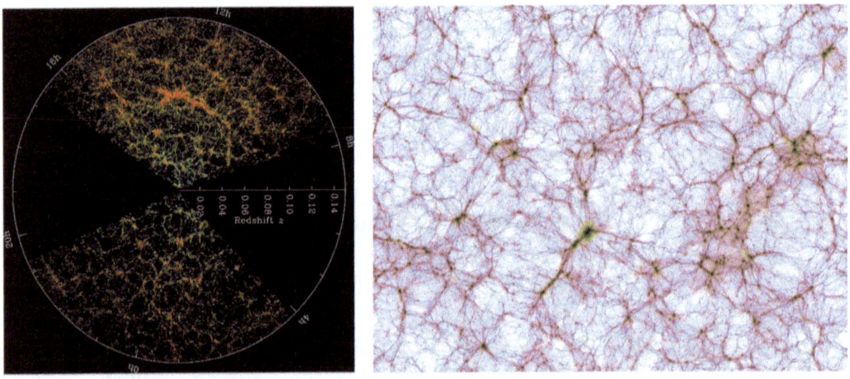

(a) (b)

Fig. 2.4 **a** A slice through the Sloan Digital Sky Survey (SDSS) galaxy map. Each dot represents a galaxy, with the Milky Way being at the centre. The region between the wedge is not mapped because the dust of our galaxy intervenes in the view of distant galaxies in these directions. The colour indicates the age of the stars in the galaxy, with the redder being made of older stars. [Credit: M. Blanton and Sloan Digital Sky Survey, sdss.org] **b** The cosmic large-scale structure as seen in a thin slice in the TNG300 simulation of IllustrisTNG. The image brightness indicates the projected baryonic mass density and the colour hue reflects the mean projected gas temperature. The width of the box here is around 1.2 billion lightyears. The underlying calculations of TNG300 shown here contain more than 30 billion resolution elements [54]. [Credit: TNG Collaboration, tng-project.org]

Cosmological simulations, such as the Millennium Simulation, have boosted our understanding of the origin and evolution of large-scale cosmic structures [57, 58]. Those simulations have successfully replicated the cosmic web's topology in a ΛCDM Universe, including the matter filaments on scales of several tens of megaparsecs and the isotropic homogenous matter distribution on larger scales [59]. In Fig. 2.4b, we see those large-scale structures in a slice from the TNG300 simulation of IllustrisTNG. This advanced magneto-hydrodynamic simulation contains more than 30 billion resolution elements [54].

In the next Sects. 2.3.1 and 2.3.2, we introduce the radiation content of the IGM including the cosmic microwave background (CMB) and extragalactic background light (EBL). Those radiation fields are crucial for the propagation of the TeV gamma rays in the IGM as we will see in Sect. 2.4.

2.3.1 Cosmic Microwave Background

The cosmic microwave background (CMB) carried out the image of the universe when it was around 380,000 years old after the Big Bang. At that time, the temperature of the universe's plasma dropped to around 3,000 degrees Kelvin (corresponds to 0.26 eV). This temperature is well below the ionization energy of hydrogen (13.6 eV). As a result of this, the electrically charged free electrons and protons formed neutral hydrogen atoms releasing photons. Those photos decoupled from baryon since temperatures have dropped below the ionization energy of hydrogen, making the Universe transparent to radiation for the first time.

The released photons at the "recombination" formed a perfect blackbody. The temperature of this blackbody has been redshifted with the expansion of the Universe reaching a temperature of 2.73 degrees kelvin today [60]. In Fig. 2.5, we see the temperature fluctuations on the order of 100μ degrees Kelvin of the CMB measured by the Planck satellite and using SMICA semi–blind spectral–matching algorithm [61]. Even though the CMB map is largely homogenous, there is a partial anisotropy consisting of the small temperature fluctuations in the CMB blackbody radiation. Measurements of the CMB polarization anisotropies are a unique probe of the early universe history and include a wealth of cosmological information [61, 62].

2.3.2 Extragalactic Background Light

The extragalactic background light (EBL) is the diffusive radiation of the light emitted across the full electromagnetic spectrum throughout the history of the universe. The EBL spans the whole electromagnetic spectrum from radio to gamma rays, however, sometimes it is referred to as the extragalactic intensity spectrum from ultraviolet to infrared [89]. The EBL spectrum includes the light emitted by all the stars, galaxies, and AGNs throughout cosmic history along with the cosmological

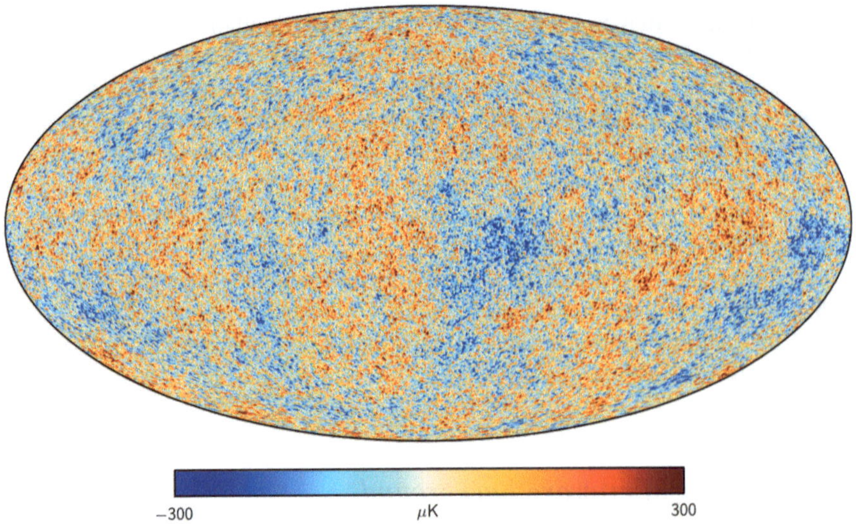

Fig. 2.5 All-sky map of the cosmic microwave background (CMB) temperature fluctuations by the Planck satellite. [Credit: ESA and the Planck Collaboration, cosmos.esa.int]

background singles of primordial phenomena, like the CMB discussed in the previous Sect. 2.3.1. The EBL might also encompass extended and diffused other signals, such as high-energy photons due to the decay or annihilation of dark matter particles [90].

The EBL measurements are due to both the ground and space-based observations. Direct absolute intensity measurements require accounting for a variety of foreground contributions both within the Solar system and our galaxy, including Zodiacal light within the Solar system and Galactic emission spanning radio, infrared, X-ray, and gamma rays [87]. This makes those measurements challenging, especially in the wavelength range of around one micrometre due to the predominant Zodiacal light contribution in this regime. However, it is possible to find lower limits on the EBL flux in this range by galaxy count surveys [91].

Indirect measurements of the EBL spectrum in the range of ultraviolet to infrared wavelengths are possible by using the absorbed TeV spectra of distant sources such as blazars and AGNs [92, 93]. The annihilation of the high-energy gamma rays with the EBL low-energy photons in the range of 0.1 μm–100 μm in wavelength produces an electron-positron pair suppressing the high-energy gamma-ray flux. Using this fundamental standard model process, one can infer the number density of the EBL photons needed to attenuate the intrinsic spectra to the absorbed one. Even though this indirect technique provides us with the best measurements in this range of the EBL, they still need improvements in the uncertainties on the intrinsic spectra of the TeV sources. In Fig. 2.6, we see the measurements of the EBL spectrum in the range of 0.1 μm–1000 μm wavelengths.

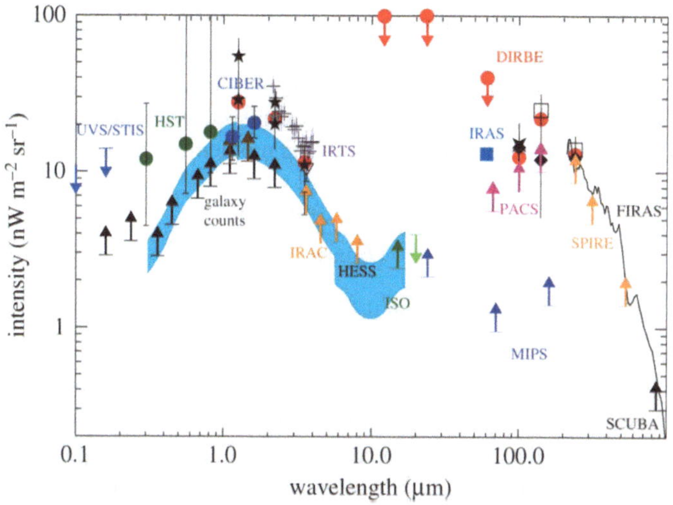

Fig. 2.6 Intensity measurements of the cosmic optical and infrared background light between 0.1 μm and 1000 μm wavelengths. Lower limits are from the integrated galaxy light from galaxy counts using SCUBA [63], Herschel/SPIRE [64], Herschel/PACS [65], Spitzer/MIPS [66, 67], ISO [68], Spitzer/IRAC [69] and Hubble [70, 71]. The data points are based on the absolute photometry estimates using IRTS (purple crosses; [72]), UVS/STIS (blue upper limits; [73, 74]), FIRAS (black line: [75, 76]), DIRBE (red circles: [77, 78], stars: [79, 80], open squares: [81]), CIBER (blue circles: [82]), Spitzer (open triangle: [83]), IRAS (bluesquare: [84]) and Hubble (green circles: [85]). The blue-shaded region is the estimate using absorption spectra by the EBL of the brightest blazars observed with H.E.S.S. [86]. Figure was taken from Cooray [87] and it is based on the figure by [88]

2.4 Blazar-induced Pair Beam Cascade

As we described in Sect. 2.2, gamma-rays with energies higher than 100 GeV have been detected from many blazars by gamma-ray telescopes like *Fermi*-LAT, VERITAS, H.E.S.S. and MAGIC. Some of those blazars are more than a billion light years away from Earth, such as the blazar of High-Frequency-Peaked BL Lacertae type (HBL) 1ES 0229+200 (with $z = 0.14$) [94]. The observations of gamma rays from such distant distances give us a unique prob of the intergalactic medium and its radiation content (EBL) [95].

As we mentioned in Sect. 2.3.2, the high-energy gamma rays propagating in the intergalactic medium interact with EBL photons producing electron-positron pairs, $\gamma + \gamma_{\rm EBL} \rightarrow e^- + e^+$. However, in order for this process to happen the two interacting photons' energy in the centre of the momentum frame has to exceed the mass of the electron and positron, leading to the following threshold condition [96]

$$E_\gamma \epsilon_{\rm EBL} \geq \frac{2(m_e c^2)^2}{1 - \cos\theta}, \tag{2.1}$$

where E_γ is the gamma-ray photon energy and $\epsilon_{\rm EBL}$ is the EBL background photon energy both in the comoving cosmological frame. Here m_e is the mass of the electron and θ is the angle between the momenta of the two interacting photons.

The pair production cross-section, $\sigma_{\gamma\gamma}$ is expressed as a function of the produced electron (positron) velocity in the centre of the momentum frame, β^*,

$$\beta^{*2} = 1 - \frac{2(m_e c^2)^2}{\left(E_\gamma \epsilon_{\rm EBL}\,(1-\cos\theta)\right)}. \tag{2.2}$$

Since we have an isotropic radiation field then averaging over $\cos\theta$ yields zero and we find the maximum cross-section to be $\sigma_{\gamma\gamma} \approx 0.25\sigma_T$ at $\epsilon_{\rm EBL} \approx \left(\frac{\rm TeV}{E_\gamma}\right)$ eV, where σ_T is the Thomson cross-section. As a consequence, the high-energy TeV photons interact most efficiently with the infrared photons where the produced electron and positron have half of the TeV photon energy each. The produced pairs tend to move in the direction of the original TeV photon with production angles of around γ^{-1}, where $\gamma = E_\gamma/2m_e c^2$ is the pairs' Lorentz factor that is of typical values between 10^4 and 10^8 [97, 98].

The exact mean free path of the high-energy gamma rays is dependent on the evolving photon density of the EBL [99], which has uncertainties in the relevant wavelengths of optical and infrared as we saw in Sect. 2.3.2. However, following Neronov and Semikoz [100] we can approximate the mean free path length for a gamma-ray with energy E_γ by

$$\lambda_{\gamma\gamma} \sim 80\,(1+z)^{-\xi}\left(\frac{E_\gamma}{10\mathrm{TeV}}\right)^{-1} \mathrm{Mpc}, \tag{2.3}$$

where $\xi = 4.5$ and $\xi = 0$ for redshit of $z \leq 1$ and $z > 1$ respectively. Typically, gamma rays with energies of TeV have mean free paths of the order of hundreds of Mpc producing the pair beams in the IGM cosmic voids.

The produced electrons and positrons in the IGM can undergo inverse Compton scattering on the CMB photons. For pairs with a Lorentz factor of about 10^6 the inverse Compton scattering on the CMB photons happens in the Thomson regime, where the pairs give a small fraction of their energy every single scattering [102]. In this regime, the mean free path for pairs with a Lorentz factor γ is [10]

$$\lambda_{\rm IC} = \frac{3 m_e c^2}{4\sigma_T U_{\rm CMB}\gamma} \approx 0.7\mathrm{Mpc}\left(\frac{E_\gamma}{\mathrm{TeV}}\right)^{-1}, \tag{2.4}$$

where $U_{\rm CMB} = 0.26$ eV cm^{-3} is the CMB integrated energy density [60]. The typical energies of the produced photons after the scattering are approximately $\epsilon_{\rm IC} = \frac{4}{3}\epsilon_{\rm CMB}\gamma^2 \approx 0.8\mathrm{GeV}\left(E_\gamma/\mathrm{TeV}\right)^2$, where the average CMB photon energy is $\epsilon_{\rm CMB} = 630\mu$eV [60]. Therefore, a primary TeV gamma ray is expected to give a secondary IC cascade in the GeV energy band.

In a nutshell, TeV primary gamma rays from distant blazars would generate a GeV-scale secondary electromagnetic cascade due to their attenuation in the IGM. This GeV secondary emission is expected to be detected in the blazar spectrum along with the survived primary emission [103–106]. However, Fermi-LAT observations at the GeV energy range of several blazars have shown that **the secondary GeV cascade is suppressed** [107, 108]. In other words, the observed GeV fluxes of blazars are significantly lower than the predicted fluxes due to the full electromagnetic cascade. This missing GeV cascade is attributed at the moment to two different physical scenarios: (1) Deflection of the pair beams by Intergalactic magnetic field (IGMF), or (2) energy loss of the pair beam by beam-plasma instability before IC cooling.

In the first scenario, the electron-positron pairs are deflected in the intergalactic magnetic fields (IGMF) along the line of sight. Sufficiently large IGMF strength would broaden the pair beam so that the produced secondary gamma-ray cascade is not detectable within the point spread function of our telescopes [103–105, 108, 109]. The secondary cascade could arrive with a time delay compared to the primary gamma-ray emission due to the extra flight distance resulting from the IGMF deflection (see Fig. 2.7). The cascade also could form a halo of extended gamma-ray emission around the point-like source. These effects have yielded lower limits constraints on the IGMF strengths for a given magnetic field correlation length. The current lower limits are set on the level of femto-Gauss magnetic field strength for a coherence length greater than 10 kpc [110].

If the magnetic field is strong enough to deflect the pair beams by a radian or more, then the cascade emission from AGNi with oblique jets (jets more than 30o off of our

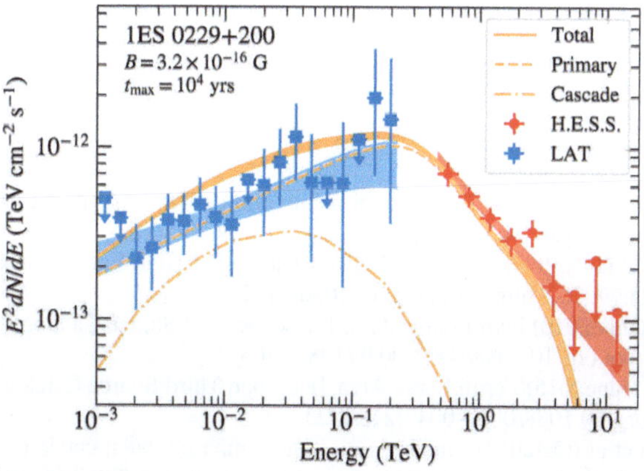

Fig. 2.7 Suppression of the cascade emission from the TeV blazar 1ES 0229+200 by the IGMF. The suppressed cascade emission is depicted as a dashed–dotted orange line, while the Observed Fermi-LAT spectra of 1ES 0229+200 in the GeV band is represented by the blue markers and lines. H.E.S.S. data points (red markers and lines) shows the high energy spectra. The primary absorbed blazar spectrum is indicated by the dashed orange line, and the combined spectrum resulting from both the secondary cascade and primary absorbed emission is illustrated by the orange solid band. Figure was taken from Aharonian et al. [101]

line of sight) should become visible [111], but the corresponding emission has not been detected in the stacking analysis of gamma-ray images in the direction of those AGNi [112–114]. This, if confirmed, suggests that the inverse Compton cooling might be ruled out as the dominant energy loss mechanism of the pair beam. An alternative energy loss mechanism of the pair beam is the beam-plasma instabilities that could drain the pair beam energy faster than the IC cooling as proposed by Broderick et al. [115].

The collective plasma effects of the TeV low-density pair beam propagating through much denser unmagnetized IGM hot background plasma have been investigated by many authors [97, 98, 115–127]. The unstable resonant plasma waves amplitude could grow exponentially taking the beam kinetic energy and eventually transferring it into a heat in the IGM plasma. If the beam energy loss time due to the instability growth is much faster than the beam cooling time by the IC scattering then the secondary gamma-ray cascade would be suppressed affecting the IGMF lower limit constraints.

However, the instability energy loss efficiency depends largely on the non-linear evolution of the instability. The non-linear evolution of the beam-plasma instabilities and the overall impact on the electromagnetic cascade are still unclear in the literature. We will discuss the non-linear evolution problem in depth in Chap. 3. If the beam's energy loss length due to the instability is smaller than the IC mean free path, then the GeV-cascade is suppressed and the beam energy is efficiently transferred to a heating of the IGM plasma. This leads to significant heating of the IGM background plasma. Cosmological simulations including this heating process can successfully reproduce the observed IGM temperature and the effective optical depth as a function of redshift as well as several other observations [128, 129].

References

1. Fichtel et al (October 1994) The First Energetic Gamma-Ray Experiment Telescope (EGRET) Source Catalog. ApJS 94:551. https://doi.org/10.1086/192082
2. Hartman et al (July1999) The Third EGRET Catalog of High-Energy Gamma-Ray Sources. ApJS 123(1):79–202. https://doi.org/10.1086/313231
3. Abdo et al (June2010) Fermi Large Area Telescope First Source Catalog. ApJS 188(2):405–436. https://doi.org/10.1088/0067-0049/188/2/405
4. Acero et al (June2015) Fermi Large Area Telescope Third Source Catalog. ApJS 218(2):23. https://doi.org/10.1088/0067-0049/218/2/23
5. Ackermann et al (Oct2013) The first fermi-lat gamma-ray burst catalog. The Astrophysical Journal Supplement Series 209(1):11. https://doi.org/10.1088/0067-0049/209/1/11
6. Abdollahi et al (Mar2020) Fermi large area telescope fourth source catalog. The Astrophysical Journal Supplement Series 247(1):33. https://doi.org/10.3847/1538-4365/ab6bcb
7. Baldini et al. Catalog of long-term transient sources in the first 10 yr of fermi-lat data. *The Astrophysical Journal Supplement Series*, 256(1):13, sep 2021. https://doi.org/10.3847/1538-4365/ac072a
8. Ajello et al. A decade of gamma-ray bursts observed by fermi-lat: The second grb catalog. *The Astrophysical Journal*, 878(1):52, jun 2019. https://doi.org/10.3847/1538-4357/ab1d4e

9. S. P. Wakely and D. Horan. TeVCat: An online catalog for Very High Energy Gamma-Ray Astronomy. In *International Cosmic Ray Conference*, volume 3 of *International Cosmic Ray Conference*, pages 1341–1344, January 2008
10. Biteau Jonathan, Meyer Manuel (2022) Gamma-Ray Cosmology and Tests of Fundamental Physics. Galaxies 10(2):39. https://doi.org/10.3390/galaxies10020039
11. Gabriele Ghisellini. *Radiative Processes in High Energy Astrophysics*, volume 873. 2013. https://doi.org/10.1007/978-3-319-00612-3
12. Giuseppe Di Sciascio. Ground-based Gamma-Ray Astronomy: an Introduction. In *Journal of Physics Conference Series*, volume 1263 of *Journal of Physics Conference Series*, page 012003, June 2019. https://doi.org/10.1088/1742-6596/1263/1/012003
13. David J. Thompson. Space detectors for gamma rays (100 mev–100 gev): From egret to fermi lat. *Comptes Rendus Physique*, 16(6):600–609, 2015. ISSN 1631-0705. https://doi.org/10.1016/j.crhy.2015.07.002. URL https://www.sciencedirect.com/science/article/pii/S1631070515001280. Gamma-ray astronomy / Astronomie des rayons gamma
14. Miller L, Peacock JA, Mead ARG (1990) The bimodal radio luminosity function of quasars. Monthly Notices of the Royal Astronomical Society 244:207–213
15. Wilson AS, Colbert EJM (January 1995) The Difference between Radio-loud and Radio-quiet Active Galaxies. ApJ 438:62. https://doi.org/10.1086/175054
16. C. Megan Urry and Paolo Padovani. Unified schemes for radio-loud active galactic nuclei. *Publications of the Astronomical Society of the Pacific*, 107(715):803, sep 1995. https://doi.org/10.1086/133630
17. Antonucci Robert (1993) Unified models for active galactic nuclei and quasars. Annual Review of Astronomy and Astrophysics 31(1):473–521. https://doi.org/10.1146/annurev.aa.31.090193.002353
18. Volker Beckmann and Chris R. Shrader. *Active Galactic Nuclei*. 2012
19. Ryan C. Hickox, Christine Jones, William R. Forman, Stephen S. Murray, Christopher S. Kochanek, Daniel Eisenstein, Buell T. Jannuzi, Arjun Dey, Michael J. I. Brown, Daniel Stern, Peter R. Eisenhardt, Varoujan Gorjian, Mark Brodwin, Ramesh Narayan, Richard J. Cool, Almus Kenter, Nelson Caldwell, and Michael E. Anderson. Host galaxies, clustering, eddington ratios, and evolution of radio, x-ray, and infrared-selected agns. *The Astrophysical Journal*, 696(1):891, apr 2009. https://doi.org/10.1088/0004-637X/696/1/891
20. Goulding et al (March 2014) Tracing the Evolution of Active Galactic Nuclei Host Galaxies over the Last 9 Gyr of Cosmic Time. ApJ 783(1):40. https://doi.org/10.1088/0004-637X/783/1/40
21. Gaidos et al. Extremely rapid bursts of tev photons from the active galaxy markarian 421. *Nature*, 383:319–320, 1996. https://api.semanticscholar.org/CorpusID:4327435
22. Maraschi L, Ghisellini G, Celotti A (September 1992) A Jet Model for the Gamma-Ray-emitting Blazar 3C 279. ApJ 397:L5. https://doi.org/10.1086/186531
23. Dermer Charles D, Schlickeiser Reinhard (October 1993) Model for the High-Energy Emission from Blazars. ApJ 416:458. https://doi.org/10.1086/173251
24. Sikora Marek, Begelman Mitchell C, Rees Martin J (January 1994) Comptonization of Diffuse Ambient Radiation by a Relativistic Jet: The Source of Gamma Rays from Blazars? ApJ 421:153. https://doi.org/10.1086/173633
25. Levinson Amir, Blandford Roger (August 1995) Pair Cascades in Extragalactic Jets. II. The Beamed X-Ray Spectrum. ApJ 449:86. https://doi.org/10.1086/176034
26. A. Mastichiadis, M. Petropoulou, and S. Dimitrakoudis. Mrk 421 as a case study for TeV and X-ray variability in leptohadronic models. *Monthly Notices of the Royal Astronomical Society*, 434(3):2684–2695, 07 2013. ISSN 0035-8711. https://doi.org/10.1093/mnras/stt1210
27. Hélène Sol and Andreas Zech. Blazars at very high energies: Emission modelling. *Galaxies*, 10(6), 2022. ISSN 2075-4434. https://doi.org/10.3390/galaxies10060105. URL https://www.mdpi.com/2075-4434/10/6/105
28. M. Cerruti, A. Zech, C. Boisson, and S. Inoue. A hadronic origin for ultra-high-frequency-peaked BL Lac objects. *Monthly Notices of the Royal Astronomical Society*, 448(1):910–927, 02 2015. ISSN 0035-8711. https://doi.org/10.1093/mnras/stu2691

29. K. Mannheim. The proton blazar. *A&A*, 269:67–76, March 1993. https://doi.org/10.48550/arXiv.astro-ph/9302006
30. G. Ghisellini, C. Righi, L. Costamante, and F. Tavecchio. The Fermi blazar sequence. *Monthly Notices of the Royal Astronomical Society*, 469(1):255–266, 04 2017. ISSN 0035-8711. https://doi.org/10.1093/mnras/stx806
31. Piero Madau. The intergalactic medium. 5 2000
32. Philip Chang, Avery E. Broderick, and Christoph Pfrommer. The cosmological impact of luminous tev blazars. ii. rewriting the thermal history of the intergalactic medium. *The Astrophysical Journal*, 752(1):23, may 2012. https://doi.org/10.1088/0004-637X/752/1/23
33. McQuinn Matthew (2016) The Evolution of the Intergalactic Medium. Ann. Rev. Astron. Astrophys. 54:313–362. https://doi.org/10.1146/annurev-astro-082214-122355
34. S. Naoz, S. Noter, and R. Barkana. The first stars in the Universe. *Monthly Notices of the Royal Astronomical Society: Letters*, 373(1):L98–L102, 11 2006. ISSN 1745-3925. https://doi.org/10.1111/j.1745-3933.2006.00251.x
35. Trenti Michele, Stiavelli Massimo (Mar2009) Formation rates of population iii stars and chemical enrichment of halos during the reionization era. The Astrophysical Journal 694(2):879. https://doi.org/10.1088/0004-637X/694/2/879
36. Bromm Volker, Larson Richard B (2004) The first stars. Annual Review of Astronomy and Astrophysics 42(1):79–118. https://doi.org/10.1146/annurev.astro.42.053102.134034
37. Bromm Volker, Kudritzki Rolf P, Loeb Abraham (May2001) Generic Spectrum and Ionization Efficiency of a Heavy Initial Mass Function for the First Stars. ApJ 552(2):464–472. https://doi.org/10.1086/320549
38. Schneider R, Ferrara A, Natarajan P, Omukai K (May2002) First stars, very massive black holes, and metals. The Astrophysical Journal 571(1):30. https://doi.org/10.1086/339917
39. Madau Piero, Meiksin Avery, Rees Martin J (February 1997) 21 Centimeter Tomography of the Intergalactic Medium at High Redshift. ApJ 475(2):429–444. https://doi.org/10.1086/303549
40. Steven R. Furlanetto, S. Peng Oh, and Frank H. Briggs. Cosmology at low frequencies: The 21cm transition and the high-redshift universe. *Physics Reports*, 433(4):181–301, 2006. ISSN 0370-1573. https://doi.org/10.1016/j.physrep.2006.08.002. URL https://www.sciencedirect.com/science/article/pii/S0370157306002730
41. Vogt et al. HIRES: the high-resolution echelle spectrometer on the Keck 10-m Telescope. In David L. Crawford and Eric R. Craine, editors, *Instrumentation in Astronomy VIII*, volume 2198, pages 362 – 375. International Society for Optics and Photonics, SPIE, 1994. https://doi.org/10.1117/12.176725
42. Esther M. Hu, Tae-Sun Kim, Lennox L. Cowie, Antoinette Songaila, and Michael Rauch. The Distribution of Column Densities and B Values in the Lyman-Alpha Forest. *AJ*, 110:1526, October 1995. https://doi.org/10.1086/117625
43. David Kirkman and David Tytler. Intrinsic properties of the $\langle z \rangle = 2.7$ lyα forest from keck spectra of quasar hs 1946+7658. *The Astrophysical Journal*, 484(2):672, aug 1997. https://doi.org/10.1086/304371
44. J. M. O'Meara, N. Lehner, J. C. Howk, J. X. Prochaska, A. J. Fox, M. A. Swain, C. R. Gelino, G. B. Berriman, and H. Tran. The First Data Release of the KODIAQ Survey. *AJ*, 150(4):111, October 2015. https://doi.org/10.1088/0004-6256/150/4/111
45. Lee et al (Feb2013) The boss lyα forest sample from sdss data release 9. The Astronomical Journal 145(3):69. https://doi.org/10.1088/0004-6256/145/3/69
46. Joop Schaye, Tom Theuns, Michael Rauch, George Efstathiou, and Wallace L.W. Sargent. The thermal history of the intergalactic medium. *Monthly Notices of the Royal Astronomical Society*, 318(3):817–826, 11 2000. ISSN 0035-8711. https://doi.org/10.1046/j.1365-8711.2000.03815.x
47. Massimo Ricotti, Nickolay Y. Gnedin, and J. Michael Shull. The evolution of the effective equation of state of the intergalactic medium. *The Astrophysical Journal*, 534(1):41, may 2000. https://doi.org/10.1086/308733

48. Faucher-Giguère Claude-André, Lidz Adam, Zaldarriaga Matias, Hernquist Lars (October 2009) A New Calculation of the Ionizing Background Spectrum and the Effects of He II Reionization. ApJ 703(2):1416–1443. https://doi.org/10.1088/0004-637X/703/2/1416
49. George D. Becker, James S. Bolton, Martin G. Haehnelt, and Wallace L. W. Sargent. Detection of extended He II reionization in the temperature evolution of the intergalactic medium. *MNRAS*, 410(2):1096–1112, January 2011. https://doi.org/10.1111/j.1365-2966.2010.17507.x
50. James S. Bolton, George D. Becker, Martin G. Haehnelt, and Matteo Viel. A consistent determination of the temperature of the intergalactic medium at redshift $\langle z \rangle = 2.4$. *Monthly Notices of the Royal Astronomical Society*, 438(3):2499–2507, 01 2014. ISSN 0035-8711. https://doi.org/10.1093/mnras/stt2374
51. Elisa Boera, Michael T. Murphy, George D. Becker, and James S. Bolton. The thermal history of the intergalactic medium down to redshift z = 1.5: a new curvature measurement. *Monthly Notices of the Royal Astronomical Society*, 441(3):1916–1933, 05 2014. ISSN 0035-8711. https://doi.org/10.1093/mnras/stu660
52. Robert A. Simcoe, Wallace L. W. Sargent, and Michael Rauch. The distribution of metallicity in the intergalactic medium at z 2.5: O vi and c iv absorption in the spectra of seven qsos*. *The Astrophysical Journal*, 606(1):92, may 2004. https://doi.org/10.1086/382777
53. Matthew M. Pieri and Martin G. Haehnelt. Pixel correlation searches for O vi in the Lyman α forest and the volume filling factor of metals in the intergalactic medium at z $\sim$ 2-3.5. *Monthly Notices of the Royal Astronomical Society*, 347(3):985–993, 01 2004. ISSN 0035-8711. https://doi.org/10.1111/j.1365-2966.2004.07278.x
54. Dylan Nelson et al. The IllustrisTNG Simulations: Public Data Release. 12 2018
55. J. Richard Bond, Lev Kofman, and Dmitry Pogosyan. How filaments of galaxies are woven into the cosmic web. *Nature*, 380(6575):603–606, April 1996. https://doi.org/10.1038/380603a0
56. Abazajian et al. The First Data Release of the Sloan Digital Sky Survey. *AJ*, 126(4):2081–2086, October 2003. https://doi.org/10.1086/378165
57. Springel Volker et al (2005) Simulating the joint evolution of quasars, galaxies and their large-scale distribution. Nature 435:629–636. https://doi.org/10.1038/nature03597
58. Mark Vogelsberger, Shy Genel, Volker Springel, Paul Torrey, Debora Sijacki, Dandan Xu, Greg Snyder, Dylan Nelson, and Lars Hernquist. Introducing the Illustris Project: simulating the coevolution of dark and visible matter in the Universe. *Monthly Notices of the Royal Astronomical Society*, 444(2):1518–1547, 08 2014. ISSN 0035-8711. https://doi.org/10.1093/mnras/stu1536
59. Libeskind et al. Tracing the cosmic web. *Monthly Notices of the Royal Astronomical Society*, 473(1):1195–1217, 08 2017. ISSN 0035-8711. https://doi.org/10.1093/mnras/stx1976
60. Fixsen DJ (Nov2009) The temperature of the cosmic microwave background. The Astrophysical Journal 707(2):916. https://doi.org/10.1088/0004-637X/707/2/916
61. Planck Collaboration, R. Adam, P. A. R. Ade, N. Aghanim, Y. Akrami, and M. I. R. Alves et al. Planck 2015 results. I. Overview of products and scientific results. *A&A*, 594:A1, September 2016. https://doi.org/10.1051/0004-6361/201527101
62. Bucher Martin (2015) Physics of the cosmic microwave background anisotropy. Int. J. Mod. Phys. D 24(02):1530004. https://doi.org/10.1142/S0218271815300049
63. Ian Smail, R. J. Ivison, A. W. Blain, and J.-P. Kneib. The nature of faint submillimetre-selected galaxies. *Monthly Notices of the Royal Astronomical Society*, 331(2):495–520, 03 2002. ISSN 0035-8711. https://doi.org/10.1046/j.1365-8711.2002.05203.x
64. Béthermin et al (June2012) HerMES: deep number counts at 250 μm, 350 μm and 500 μm in the COSMOS and GOODS-N fields and the build-up of the cosmic infrared background. A&A 542:A58. https://doi.org/10.1051/0004-6361/201118698
65. Berta et al (July2010) Dissecting the cosmic infra-red background with Herschel/PEP. A&A 518:L30. https://doi.org/10.1051/0004-6361/201014610
66. Dole et al. Far-infrared source counts at 70 and 160 microns in spitzer deep surveys. *The Astrophysical Journal Supplement Series*, 154(1):87, sep 2004. https://doi.org/10.1086/422472

67. Papovich et al. The 24 micron source counts in deep spitzer space telescope surveys*. *The Astrophysical Journal Supplement Series*, 154(1):70, sep 2004. https://doi.org/10.1086/422880
68. D. Elbaz, C. J. Cesarsky, D. Fadda, H. Aussel, F. X. Désert, A. Franceschini, H. Flores, M. Harwit, J. L. Puget, J. L. Starck, D. L. Clements, L. Danese, D. C. Koo, and R. Mandolesi. Source counts from the 15 mu m ISOCAM Deep Surveys. *A&A*, 351:L37–L40, November 1999. https://doi.org/10.48550/arXiv.astro-ph/9910406
69. G. G. Fazio, M. L. N. Ashby, P. Barmby, J. L. Hora, J.-S. Huang, M. A. Pahre, Z. Wang, S. P. Willner, R. G. Arendt, S. H. Moseley, M. Brodwin, P. Eisenhardt, Daniel Stern, E. V. Tollestrup, and E. L. Wright. Number counts at 3 μm & $<\lambda<10\mu$m from the spitzer space telescope. *The Astrophysical Journal Supplement Series*, 154(1):39, sep 2004. https://doi.org/10.1086/422585
70. Gardner Jonathan P, Brown Thomas M, Ferguson Henry C (October 2000) Ultraviolet Galaxy Counts from Space Telescope Imaging Spectrograph Observations of the Hubble Deep Fields. ApJ 542(2):L79–L82. https://doi.org/10.1086/312930
71. Madau Piero, Pozzetti Lucia (February 2000) Deep galaxy counts, extragalactic background light and the stellar baryon budget. MNRAS 312(2):L9–L15. https://doi.org/10.1046/j.1365-8711.2000.03268.x
72. Matsumoto T, Matsuura S, Murakami H, Tanaka M, Freund M, Lim M, Cohen M, Kawada M, Noda M (June2005) Infrared Telescope in Space Observations of the Near-Infrared Extragalactic Background Light. ApJ 626(1):31–43. https://doi.org/10.1086/429383
73. Edelstein Jerry, Bowyer Stuart, Lampton Michael (August 2000) Reanalysis ofVoyager Ultraviolet Spectrometer Limits to the Extreme-Ultraviolet and Far-Ultraviolet Diffuse Astronomical Flux. ApJ 539(1):187–190. https://doi.org/10.1086/309192
74. Thomas M. Brown, Randy A. Kimble, Henry C. Ferguson, Jonathan P. Gardner, Nicholas R. Collins, and Robert S. Hill. Measurements of the Diffuse Ultraviolet Background and the Terrestrial Airglow with the Space Telescope Imaging Spectrograph. *AJ*, 120(2):1153–1159, August 2000. https://doi.org/10.1086/301468
75. G. Lagache, L. M. Haffner, R. J. Reynolds, and S. L. Tufte. Evidence for dust emission in the Warm Ionised Medium using WHAM data. *A&A*, 354:247–252, February 2000. https://doi.org/10.48550/arXiv.astro-ph/9911355
76. Fixsen DJ, Dwek E, Mather JC, Bennett CL, Shafer RA (November 1998) The Spectrum of the Extragalactic Far-Infrared Background from the COBE FIRAS Observations. ApJ 508(1):123–128. https://doi.org/10.1086/306383
77. Edward L. Wright. DIRBE minus 2MASS: Confirming the Cosmic Infrared Background at 2.2 Microns. *ApJ*, 553(2):538–544, June 2001. https://doi.org/10.1086/320942
78. Wright EL (April 2004) COBE observations of the cosmic infrared background. New A Rev. 48(5–6):465–468. https://doi.org/10.1016/j.newar.2003.12.054
79. L. Cambrésy, W. T. Reach, C. A. Beichman, and T. H. Jarrett. The Cosmic Infrared Background at 1.25 and 2.2 Microns Using DIRBE and 2MASS: A Contribution Not Due to Galaxies? *ApJ*, 555(2):563–571, July 2001. https://doi.org/10.1086/321470
80. L. R. Levenson, E. L. Wright, and B. D. Johnson. DIRBE Minus 2MASS: Confirming the CIRB in 40 New Regions at 2.2 and 3.5 μm. *ApJ*, 666(1):34–44, September 2007. https://doi.org/10.1086/520112
81. Hauser MG, Arendt RG, Kelsall T, Dwek E, Odegard N, Weiland JL, Freudenreich HT, Reach WT, Silverberg RF, Moseley SH, Pei YC, Lubin P, Mather JC, Shafer RA, Smoot GF, Weiss R, Wilkinson DT, Wright EL (November 1998) The COBE Diffuse Infrared Background Experiment Search for the Cosmic Infrared Background. I. Limits and Detections. ApJ 508(1):25–43. https://doi.org/10.1086/306379
82. Michael Zemcov, Joseph Smidt, Toshiaki Arai, James Bock, Asantha Cooray, Yan Gong, Min Gyu Kim, Phillip Korngut, Anson Lam, Dae Hee Lee, Toshio Matsumoto, Shuji Matsuura, Uk Won Nam, Gael Roudier, Kohji Tsumura, and Takehiko Wada. On the origin of near-infrared extragalactic background light anisotropy. *Science*, 346(6210):732–735, November 2014. https://doi.org/10.1126/science.1258168

83. L. R. Levenson and E. L. Wright. Probing the 3.6 μm CIRB with Spitzer in Three DIRBE Dark Spots. *ApJ*, 683(2):585–596, August 2008. https://doi.org/10.1086/589808
84. Miville-Deschênes MA, Lagache G, Puget JL (October 2002) Power spectrum of the cosmic infrared background at 60 and 100 IRAS. A&A 393:749–756. https://doi.org/10.1051/0004-6361:20020929
85. Bernstein Rebecca A (September 2007) The Optical Extragalactic Background Light: Revisions and Further Comments. ApJ 666(2):663–673. https://doi.org/10.1086/519824
86. H. E. S. S. Collaboration, A. Abramowski, F. Acero, F. Aharonian, A. G. Akhperjanian, and G. Anton et al. Measurement of the extragalactic background light imprint on the spectra of the brightest blazars observed with H.E.S.S. *A&A*, 550:A4, February 2013. https://doi.org/10.1051/0004-6361/201220355
87. Cooray Asantha (March 2016) Extragalactic background light measurements and applications. Royal Society Open Science 3(3):150555. https://doi.org/10.1098/rsos.150555
88. H. Dole, G. Lagache, J. L. Puget, K. I. Caputi, N. Fernández-Conde, E. Le Floc'h, C. Papovich, P. G. Pérez-González, G. H. Rieke, and M. Blaylock. The cosmic infrared background resolved by Spitzer. Contributions of mid-infrared galaxies to the far-infrared background. *A&A*, 451(2):417–429, May 2006. https://doi.org/10.1051/0004-6361:20054446
89. Eli Dwek and Frank Krennrich. The extragalactic background light and the gamma-ray opacity of the universe. *Astroparticle Physics*, 43:112–133, 2013. ISSN 0927-6505. https://doi.org/10.1016/j.astropartphys.2012.09.003. URL https://www.sciencedirect.com/science/article/pii/S0927650512001752. Seeing the High-Energy Universe with the Cherenkov Telescope Array - The Science Explored with the CTA
90. Abdo et al (April 2010) Constraints on cosmological dark matter annihilation from the Fermi-LAT isotropic diffuse gamma-ray measurement. J. Cosmology Astropart. Phys. 2010(4):014. https://doi.org/10.1088/1475-7516/2010/04/014
91. Piero Madau and Lucia Pozzetti. Deep galaxy counts, extragalactic background light and the stellar baryon budget. *Monthly Notices of the Royal Astronomical Society*, 312(2):L9–L15, 02 2000. ISSN 0035-8711. https://doi.org/10.1046/j.1365-8711.2000.03268.x
92. H. E. S. S. Collaboration, H. Abdalla, A. Abramowski, F. Aharonian, F. Ait Benkhali, and A. G. Akhperjanian et al. Measurement of the EBL spectral energy distribution using the VHE γ-ray spectra of H.E.S.S. blazars. *A&A*, 606:A59, October 2017. https://doi.org/10.1051/0004-6361/201731200
93. Abeysekara et al (Nov2019) Measurement of the extragalactic background light spectral energy distribution with VERITAS. The Astrophysical Journal 885(2):150. https://doi.org/10.3847/1538-4357/ab4817
94. Aharonian et al (November 2007) New constraints on the mid-IR EBL from the HESS discovery of VHE γ-rays from 1ES 0229+200. A&A 475(2):L9–L13. https://doi.org/10.1051/0004-6361:20078462
95. Alberto Franceschini. Photon–photon interactions and the opacity of the universe in gamma rays. *Universe*, 7(5), 2021. ISSN 2218-1997. https://doi.org/10.3390/universe7050146. URL https://www.mdpi.com/2218-1997/7/5/146
96. Gould Robert J, Schréder Gérard P (Mar1967) Pair production in photon-photon collisions. Phys. Rev. 155:1404–1407 10.1103/PhysRev.155.1404. URL https://link.aps.org/doi/10.1103/PhysRev.155.1404
97. Miniati Francesco, Elyiv Andrii (May2013) RELAXATION OF BLAZAR-INDUCED PAIR BEAMS IN COSMIC VOIDS. The Astrophysical Journal 770(1):54. https://doi.org/10.1088/0004-637x/770/1/54
98. Schlickeiser R, Ibscher D, Supsar M (Oct2012) PLASMA EFFECTS ON FAST PAIR BEAMS IN COSMIC VOIDS. The Astrophysical Journal 758(2):102. https://doi.org/10.1088/0004-637x/758/2/102
99. Kneiske, T. M., Bretz, T., Mannheim, K., and Hartmann, D. H. Implications of cosmological gamma-ray absorption - ii. modification of gamma-ray spectra. *A&A*, 413(3):807–815, 2004. https://doi.org/10.1051/0004-6361:20031542

100. Neronov A, Semikoz DV (Dec2009) Sensitivity of γ-ray telescopes for detection of magnetic fields in the intergalactic medium. Phys. Rev. D 80:123012. https://doi.org/10.1103/PhysRevD.80.123012
101. F. Aharonian et al. Constraints on the Intergalactic Magnetic Field Using Fermi-LAT and H.E.S.S. Blazar Observations. *Astrophys. J. Lett.*, 950(2):L16, 2023. https://doi.org/10.3847/2041-8213/acd777
102. Blumenthal George R, Gould Robert J (Apr1970) Bremsstrahlung, synchrotron radiation, and compton scattering of high-energy electrons traversing dilute gases. Rev. Mod. Phys. 42:237–270. https://doi.org/10.1103/RevModPhys.42.237
103. Aharonian FA, Coppi PS, Voelk HJ (March 1994) Very High Energy Gamma Rays from Active Galactic Nuclei: Cascading on the Cosmic Background Radiation Fields and the Formation of Pair Halos. ApJ 423:L5. https://doi.org/10.1086/187222
104. Plaga R (March 1995) Detecting intergalactic magnetic fields using time delays in pulses of γ-rays. Nature 374(6521):430–432. https://doi.org/10.1038/374430a0
105. Neronov A, Semikoz DV (December 2009) Sensitivity of γ-ray telescopes for detection of magnetic fields in the intergalactic medium. Physical Review D 80(12):123012. https://doi.org/10.1103/PhysRevD.80.123012
106. Elyiv A, Neronov A, Semikoz DV (Jul2009) Gamma-ray induced cascades and magnetic fields in the intergalactic medium. Phys. Rev. D 80:023010. https://doi.org/10.1103/PhysRevD.80.023010
107. Abdo et al (Jan2010) SPECTRAL PROPERTIES OF BRIGHTFERMI-DETECTED BLAZARS IN THE GAMMA-RAY BAND. The Astrophysical Journal 710(2):1271–1285. https://doi.org/10.1088/0004-637x/710/2/1271
108. Neronov Andrii, Vovk Ievgen (April 2010) Evidence for Strong Extragalactic Magnetic Fields from Fermi Observations of TeV Blazars. Science 328(5974):73. https://doi.org/10.1126/science.1184192
109. Taylor AM, Vovk I, Neronov A (May2011) Extragalactic magnetic fields constraints from simultaneous GeV-TeV observations of blazars. A&A 529:A144. https://doi.org/10.1051/0004-6361/201116441
110. Fermi-LAT Collaboration, F. Aharonian, J. Aschersleben, M. Backes, V. Barbosa Martins, and R. Batzofin et al. Constraints on the Intergalactic Magnetic Field Using Fermi-LAT and H.E.S.S. Blazar Observations. *ApJ*, 950(2):L16, June 2023. https://doi.org/10.3847/2041-8213/acd777
111. Broderick Avery E, Tiede Paul, Shalaby Mohamad, Pfrommer Christoph, Puchwein Ewald, Chang Philip, Lamberts Astrid (2016) Bow Ties in the Sky I: The Angular Structure of Inverse Compton Gamma-ray Halos in the Fermi Sky. Astrophys. J. 832(2):109. https://doi.org/10.3847/0004-637X/832/2/109
112. Paul Tiede, Avery E. Broderick, Mohamad Shalaby, Christoph Pfrommer, Ewald Puchwein, Philip Chang, and Astrid Lamberts. Bow Ties in the Sky. II. Searching for Gamma-Ray Halos in the Fermi Sky Using Anisotropy. *Astrophys. J.*, 850(2):157, 2017. https://doi.org/10.3847/1538-4357/aa9375
113. Broderick Avery E, Tiede Paul, Chang Philip, Lamberts Astrid, Pfrommer Christoph, Puchwein Ewald, Shalaby Mohamad, Werhahn Maria (Nov2018) Missing gamma-ray halos and the need for new physics in the gamma-ray sky. The Astrophysical Journal 868(2):87. https://doi.org/10.3847/1538-4357/aae5f2
114. M. Ackermann, M. Ajello, L. Baldini, J. Ballet, G. Barbiellini, D. Bastieri, R. Bellazzini, E. Bissaldi, R. D. Blandford, E. D. Bloom, and et al. The search for spatial extension in high-latitude sources detected by the fermi large area telescope. *The Astrophysical Journal Supplement Series*, 237(2):32, Aug 2018. ISSN 1538-4365. https://doi.org/10.3847/1538-4365/aacdf7
115. Avery E. Broderick, Philip Chang, and Christoph Pfrommer. THE COSMOLOGICAL IMPACT OF LUMINOUS TeV BLAZARS. i. IMPLICATIONS OF PLASMA INSTABILITIES FOR THE INTERGALACTIC MAGNETIC FIELD AND EXTRAGALACTIC GAMMA-RAY BACKGROUND. *The Astrophysical Journal*, 752(1):22, may 2012. https://doi.org/10.1088/0004-637x/752/1/22

116. R. Schlickeiser, S. Krakau, and M. Supsar. PLASMA EFFECTS ON FAST PAIR BEAMS. II. REACTIVE VERSUS KINETIC INSTABILITY OF PARALLEL ELECTROSTATIC WAVES. *The Astrophysical Journal*, 777(1):49, oct 2013. https://doi.org/10.1088/0004-637x/777/1/49
117. Broderick Avery E, Pfrommer Christoph, Puchwein Ewald, Chang Philip (August 2014) Implications of Plasma Beam Instabilities for the Statistics of the Fermi Hard Gamma-Ray Blazars and the Origin of the Extragalactic Gamma-Ray Background. ApJ 790(2):137. https://doi.org/10.1088/0004-637X/790/2/137
118. Sironi Lorenzo, Giannios Dimitrios (May2014) RELATIVISTIC PAIR BEAMS FROM TeV BLAZARS: A SOURCE OF REPROCESSED GeV EMISSION RATHER THAN INTERGALACTIC HEATING. The Astrophysical Journal 787(1):49. https://doi.org/10.1088/0004-637x/787/1/49
119. Chang Philip, Broderick Avery E, Pfrommer Christoph, Puchwein Ewald, Lamberts Astrid, Shalaby Mohamad (Dec2014) THE EFFECT OF NONLINEAR LANDAU DAMPING ON ULTRARELATIVISTIC BEAM PLASMA INSTABILITIES. The Astrophysical Journal 797(2):110. https://doi.org/10.1088/0004-637x/797/2/110
120. Markus Supsar and Reinhard Schlickeiser. PLASMA EFFECTS ON FAST PAIR BEAMS. III. OBLIQUE ELECTROSTATIC GROWTH RATES FOR PERPENDICULAR MAXWELLIAN PAIR BEAMS. *The Astrophysical Journal*, 783(2):96, feb 2014. https://doi.org/10.1088/0004-637x/783/2/96
121. Chang Philip, Broderick Avery E, Pfrommer Christoph, Puchwein Ewald, Lamberts Astrid, Shalaby Mohamad, Vasil Geoffrey (Dec2016) The linear instability of dilute ultrarelativistic e± pair beams. The Astrophysical Journal 833(1):118. https://doi.org/10.3847/1538-4357/833/1/118
122. Kempf A, Kilian P, Spanier F (January 2016) Energy loss in intergalactic pair beams: Particle-in-cell simulation. Astronomy & Astrophysics 585:A132. https://doi.org/10.1051/0004-6361/201527521
123. Rafighi I, Vafin S, Pohl M, Niemiec J (November 2017) Plasma effects on relativistic pair beams from TeV blazars: PIC simulations and analytical predictions. Astronomy & Astrophysics 607:A112. https://doi.org/10.1051/0004-6361/201731127
124. S. Vafin, I. Rafighi, M. Pohl, and J. Niemiec. The electrostatic instability for realistic pair distributions in blazar/ebl cascades. *The Astrophysical Journal*, 857(1):43, Apr 2018. ISSN 1538-4357. https://doi.org/10.3847/1538-4357/aab552
125. Vafin S, Deka PJ, Pohl M, Bohdan A (Feb2019) Revisit of nonlinear landau damping for electrostatic instability driven by blazar-induced pair beams. The Astrophysical Journal 873(1):10. https://doi.org/10.3847/1538-4357/ab017b
126. Rafael Alves Batista, Andrey Saveliev, and Elisabete M. de Gouveia Dal Pino. The Impact of Plasma Instabilities on the Spectra of TeV Blazars. *Mon. Not. Roy. Astron. Soc.*, 489(3):3836–3849, 2019. https://doi.org/10.1093/mnras/stz2389
127. Shalaby Mohamad, Broderick Avery E, Chang Philip, Pfrommer Christoph, Puchwein Ewald, Lamberts Astrid (2020) The growth of the longitudinal beam–plasma instability in the presence of an inhomogeneous background. Journal of Plasma Physics 86(2):535860201. https://doi.org/10.1017/S0022377820000215
128. Ewald Puchwein, Christoph Pfrommer, Volker Springel, Avery E. Broderick, and Philip Chang. The Lyman α forest in a blazar-heated Universe. *Monthly Notices of the Royal Astronomical Society*, 423(1):149–164, 05 2012a. ISSN 0035-8711. https://doi.org/10.1111/j.1365-2966.2012.20738.x
129. Lamberts Astrid, Puchwein Ewald, Pfrommer Christoph, Chang Philip, Shalaby Mohamad, Broderick Avery, Tiede Paul, Rudie Gwen (May2022) Constraining blazar heating with the $2 \lesssim z \lesssim 3$ Lyman-α forest. MNRAS 512(2):3045–3059. https://doi.org/10.1093/mnras/stac553

Chapter 3
Exploring the Foundations of Beam-Plasma Instability

In this chapter, we review the theoretical frameworks that describe beam-plasma instability. We will review the main characteristics of the intergalactic medium plasma in Sect. 3.1. The beam-plasma instability is a collective process that can be described by the plasma kinetic theory in the most general case. The kinetic theory involves the Vlasov equation of the particle distribution function coupled with the Maxwell equations and it will be introduced in Sect. 3.2. By perturbing a given equilibrium distribution with infinitesimal perturbations around the equilibrium distribution, one can find the dispersion relations that can tell us whether those perturbations decay or grow. When the imaginary component of the plasma wave frequency is positive ($\omega_i > 0$), the waves grow from the small thermal fluctuations taking the kinetic energy of the beam. This phenomenon is called beam-plasma instability and will be introduced in Sect. 3.3. The relevant beam plasma instability for the blazar-induced pair beams is the electrostatic beam-plasma instability in the kinetic regime and will be introduced in Sect. 3.4. Finally, the non-linear effects and the limitations of the linear theory will be presented in Sects. 3.5 and 3.6.

3.1 Intergalactic Medium Plasma

We have introduced the intergalactic medium (IGM) in Sect. 2.3 of the previous chapter. We saw that this medium has a very low density of around 10^{-7} particles per centimetre cube and a temperature of the order 10^4 K. The IGM plasma properties, such as its density and temperature, are crucial for the growth and evolution of the blazar-induced pair beam instabilities. In this section, we will be using those values to look at the important characteristics of the IGM plasma.

The Debye length is a characteristic scale that describes the length at which most of the particles' charge is neutralized by neighbouring plasma charges. The Debye length is defined as

M. S. A. Alawashra, *Plasma Instabilities of TeV Pair Beams Induced by Blazars*, Springer Theses, https://doi.org/10.1007/978-3-031-98568-3_3

$$\lambda_D = \sqrt{\frac{kT_e}{4\pi n_e e^2}}, \tag{3.1}$$

where T_e and n_e are the IGM temperature and density respectively. If there are many particles in the sphere with the Debye length radius then the collective effects of the plasma dominate. This condition is well satisfied for the IGM plasma.

For the IGM plasma, the mean free path of the coulomb collisions is much larger than the Debye length of the plasma. Therefore the IGM plasma can be considered as collisionless plasma where the interactions of particles are to be described by the collective influence of many particles rather than the two-body Coulomb collisions.

Another collective phenomenon besides Debye shielding is plasma oscillation, where the electrons oscillate around their motion, with the plasma frequency ω_p. Ignoring the thermal motion of the electrons, this frequency is given by

$$\omega_p = \sqrt{\frac{4\pi e^2 n_e}{m_e}}. \tag{3.2}$$

The plasma frequency is the largest frequency of the electric field fluctuations that could propagate in the IGM plasma. Those fluctuations happen in the background of the plasma equilibrium state, which changes slowly in space and time or stays constant. The equilibrium state can evolve slowly under the slowly changing external conditions or due to the average effect of many small fluctuations.

3.2 Kinetic Theory

In the context of a collisionless plasma like the IGM plasma, it is crucial to determine the specific shape of the velocity distribution function using a statistical model. The distribution function plays a central role in the analytical plasma description referred to as the kinetic treatment. The kinetic treatment forms the primary framework for analyzing plasma interactions in this thesis. It revolves around solving the Vlasov equation, coupled with Maxwell's equations, to address the relevant plasma physics problems.

Using the fact that the particle distribution function is conserved locally, the Vlasov equation describes the evolution of the particle distribution function of species s, $f_s(\mathbf{r}, \mathbf{v}, t)$, in space, velocity and time

$$\frac{df_s}{dt} = \frac{\partial f_s}{\partial t} + \mathbf{v} \cdot \frac{\partial f_s}{\partial \mathbf{r}} + \frac{\mathbf{F}}{m_s} \cdot \frac{\partial f_s}{\partial \mathbf{v}} = 0, \tag{3.3}$$

where $\mathbf{F}_s$ is the Lorentz force that describes the feedback of electric and magnetic fields on the particles

$$\mathbf{F}_s = q_s \left(\mathbf{E} + \frac{\mathbf{v} \times \mathbf{B}}{c} \right). \tag{3.4}$$

The time and spatial evolution of the electric fields $\mathbf{E}$ and the magnetic fields $\mathbf{B}$ is governed by the Maxwell's equations

$$\nabla \cdot \mathbf{E} = 4\pi \rho, \tag{3.5}$$

$$\nabla \cdot \mathbf{B} = 0, \tag{3.6}$$

$$\nabla \times \mathbf{E} + \frac{1}{c}\frac{\partial \mathbf{B}}{\partial t} = 0, \tag{3.7}$$

$$\nabla \times \mathbf{B} - \frac{1}{c}\frac{\partial \mathbf{E}}{\partial t} = \frac{4\pi}{c}\mathbf{J}, \tag{3.8}$$

where ρ is the total charge density

$$\rho(\mathbf{r}, t) = \sum_s q_s \int d\mathbf{v} f_s(\mathbf{r}, \mathbf{v}, t), \tag{3.9}$$

and $\mathbf{J}$ is the total current density

$$\mathbf{J} = \sum_s q_s \int d\mathbf{v}\mathbf{v} f(\mathbf{r}, \mathbf{v}, t). \tag{3.10}$$

This provides us with a closed system of equations describing the plasma, where the electromagnetic fields are evolved by Maxwell's equations given the total charge and current densities. Using the fields, the distribution of every species is evolved by a separate Vlasov equation, making the system of Vlasov-Maxwell equations eventually closed.

The Vlasov-Maxwell system can be solved numerically because obtaining analytical solutions is usually a challenging task. An alternative analytical approach is to assume an equilibrium state and explore the effect of small linear perturbations of the average electric, $\mathbf{E}_0$, and magnetic, $\mathbf{B}_0$, fields as well the particle distribution function, f_0,

$$\mathbf{E}(\mathbf{r}, t) = \mathbf{E}_0(\mathbf{r}) + \delta\mathbf{E}(\mathbf{r}, t), \tag{3.11}$$

$$\mathbf{B}(\mathbf{r}, t) = \mathbf{B}_0(\mathbf{r}) + \delta\mathbf{B}(\mathbf{r}, t), \tag{3.12}$$

$$f(\mathbf{r}, \mathbf{v}, t) = f_0(\mathbf{r}, \mathbf{v}) + \delta f(\mathbf{r}, \mathbf{v}, t), \tag{3.13}$$

where all the perturbations ($\delta\mathbf{E}$,$\delta\mathbf{B}$, δf) are much smaller than the equilibrium states ($\mathbf{E}_0$,$\mathbf{B}_0$,f_0).

Furthermore one can look for wavelike perturbations where

$$\delta\mathbf{E}(\mathbf{r}, t) = \mathbf{E}_1 \exp i(\mathbf{k} \cdot \mathbf{v} - \omega t), \tag{3.14}$$

$$\delta\mathbf{B}(\mathbf{r}, t) = \mathbf{B}_1 \exp i(\mathbf{k} \cdot \mathbf{v} - \omega t), \tag{3.15}$$

$$\delta f(\mathbf{r}, \mathbf{v}, t) = f_1(\mathbf{v}) \exp i(\mathbf{k} \cdot \mathbf{v} - \omega t). \tag{3.16}$$

Those perturbations can be used with the Maxwell and Vlasov equation to derive the so-called dispersion relation for the wave frequency dependence on the wave vector $\Phi = \Phi(\omega, \mathbf{k})$ for a given plasma system. Solving this implicit equation for a given plasma system provides information about the properties of waves in the system, including their phase velocity, group velocity, and temporal stability. The stability of perturbations with a given wavevector $\mathbf{k}$ is determined by the imaginary part of the frequency ω; if it is positive, the wave modes are unstable and the waves can grow in time, and if it is negative, the wave modes get damped.

3.2.1 Landau Damping

We will look at the one-dimensional case now where there are no electric fields, $\mathbf{E}_0 = 0$, and the plasma is unmagnetized, $\mathbf{B} = 0$, in electrostatic approximation. The distribution function is still the sum of the steady-state and a perturbation, $f = f_0 + \delta f$. With all the perturbations being in the form of plane waves in one-dimension approximation, we get the following equations as a first-order approximation of the Vlasov equation

$$\frac{\partial f_0}{\partial t} + v \cdot \frac{\partial f_0}{\partial x} = 0, \tag{3.17}$$

and

$$i(\omega - vk) f_1 = \frac{q}{m} E_1 \frac{\partial f_0}{\partial v}. \tag{3.18}$$

Using Gauss law, Eq. 3.5, and the total charge density, Eq. 3.9, we get

$$ikE_1 = 4\pi q \int_{-\infty}^{\infty} dv f_1. \tag{3.19}$$

Combining Eqs. 3.18 and 3.19 we get

$$1 - \frac{4\pi q^2}{mk} \int_{-\infty}^{\infty} dv \frac{\frac{\partial f_0}{\partial v}}{vk - \omega} = 0. \tag{3.20}$$

We notice that there is a singularity $vk - \omega = 0$. Therefore, there is a complex solution of the frequency $\omega = \omega_R + i\omega_i$ that can satisfy Eq. 3.20. A solution was found by Landau [1] for an equilibrium distribution in the form of the Maxwellian distribution. The final result is given for the case when $\omega_R \gg \omega_i$ by

$$\omega_R(k) = \omega_p \left(1 + \frac{3}{2}\lambda_D^2 k^2\right), \tag{3.21}$$

for the real part and

$$\omega_i(k) = -\omega_p \sqrt{\frac{\pi}{8}} \left(\frac{1}{k\lambda_D}\right)^3 \exp -\frac{1}{2}\left(\frac{1}{k\lambda_D}\right)^2, \tag{3.22}$$

for the imaginary part, where ω_p is the plasma frequency defined by Eq. 3.2 and λ_D is the Debye length defined by Eq. 3.1.

Substituting the wave frequency in the perturbed electric field Eq. 3.14, yields

$$\delta E = E_1 \exp \omega_i t \exp i(kx - \omega_R t). \tag{3.23}$$

We see that the waves are excited and amplitudes grow when $\omega_i > 0$ and decay when $\omega_i < 0$. For the case of Landau damping, the sign of ω_i is dependent on the velocity derivative of the equilibrium distribution function. So the waves grow if the distribution function increases with the velocity derivative, and they are damped once the distribution function derivative declines. The Maxwellian distribution decreases with the speed which explains the negative sign in Eq. 3.22, making the electric field associated with the waves damped, which is known as Landau damping.

3.3 Beam-Plasma Instabilities

The system of relativistic beam plasma is a very complex system with rapid non-linear interactions of the unstable waves and non-linear feedback of the unstable waves on the beam. This system is difficult to model including all the operating processes. As we introduced in the previous chapter, the blazar-induced pair beams have a very low density and propagate through a much denser background IGM plasma that is neutral and consists of ions and electrons. In this IGM background medium, we can safely assume a very weak or zero magnetic field ($\mathbf{B}_0 = 0$) and a zero large-scale electric field ($\mathbf{E}_0 = 0$).

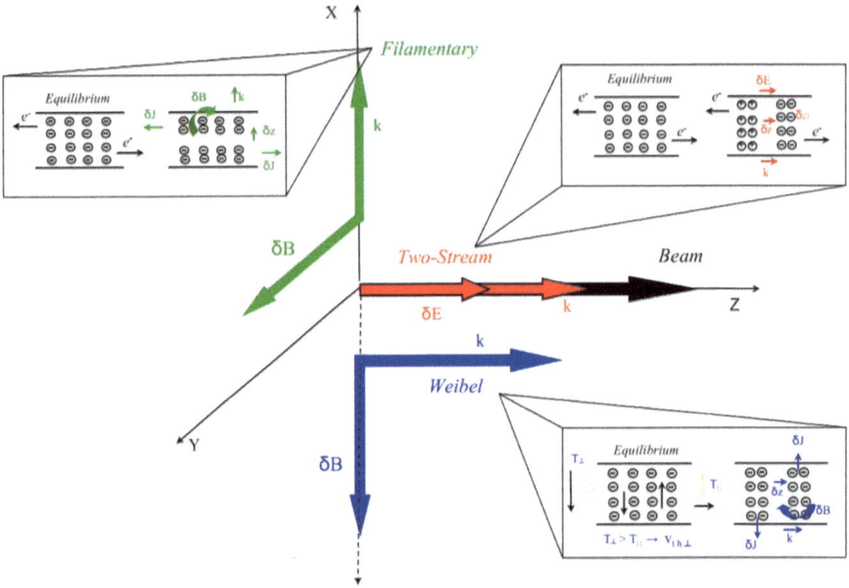

Fig. 3.1 The filamentation (green), two-stream (red) and Weibel (blue) modes. Figure was taken from Cottrill et al. [2]

One can use the dispersion relation yielded from the Vlasov-Maxwell system of the beam-plasma system to explore the relevant instabilities and their growth rates. For a cold or a monoenergetic beam, where the momentum spread of both the beam and background plasma is very small, several types of beam-plasma instabilities can develop [3]. Those instabilities include the two-stream, oblique, Weibel, and filamentary instabilities. These instabilities usually start with a linear growth phase, followed by a phase of non-linear interactions where the instability can eventually saturate at some level.

In Fig. 3.1, different types of instabilities are demonstrated. The simplest case is the so-called two-stream electrostatic instability with no magnetic fluctuations ($\mathbf{k} \times \delta\mathbf{E} = 0$ and $\delta\mathbf{B} = 0$) and wave vector $\mathbf{k}$ parallel to the beam propagation axis ($\mathbf{k}||\mathbf{v}_b$) as seen in Fig. 3.1 [4]. If the electrostatic modes have a finite angle to the beam propagation direction then they are called oblique modes. The filamentary instability is an electromagnetic instability, consisting of both electric and magnetic field fluctuations, where the wave vector is perpendicular to the beam direction ($\mathbf{k} \perp \mathbf{v}_b$) as shown in Fig. 3.1.

The origin of both the two-stream and filamentary instabilities is the counter-streaming motion of the beams [2]. If the electromagnetic instability is driven by the anisotropies in the beam temperature distribution then it's called Weibel instability [5]. The decisive quantity for the development of those different instabilities is the maximum growth rate. The growth rates of the fastest modes for the filamentation, two-stream and oblique instabilities are given by [3]

$$\frac{\omega_{i,\mathrm{F}}}{\omega_p} = \beta\sqrt{\frac{\alpha}{\gamma}}, \tag{3.24}$$

$$\frac{\omega_{i,\mathrm{TS}}}{\omega_p} = \frac{\sqrt{3}}{2^{4/3}}\frac{\alpha^{1/3}}{\gamma}, \tag{3.25}$$

$$\frac{\omega_{i,\mathrm{O}}}{\omega_p} = \frac{\sqrt{3}}{2^{4/3}}\left(\frac{\alpha}{\gamma}\right)^{1/3}, \tag{3.26}$$

where γ is the average beam particles Lorentz factor and $\alpha = n_b/n_e$ is the density ratio of the beam density n_b to the background plasma density n_e. For the typical blazar-induced TeV pair beams in the IGM, the Lorentz boost of the pair beam is of the order $\gamma = 10^3 - 10^8$ and the beam densities are $n_b = 10^{-24} - 10^{-19}\mathrm{cm}^{-3}$. With the IGM density of the order $n_e = 10^{-7}\mathrm{cm}^{-3}$, we obtain a density ratio of around $\alpha = 10^{-17} - 10^{-12}$ for the realistic blazar-induced pair beams.

Rafighi et al. [6] demonstrated using an analytical model that the condition for the Weibel modes to grow is

$$\alpha > \left(2 - \frac{\pi}{2}\right)\frac{k_B T_b}{\gamma m_e c^2}, \tag{3.27}$$

where T_b is the beam temperature in its rest frame. For the realistic blazar-induced pair beams, the beam temperature is around $k_B T_b = m_e c^2$ and $\alpha\gamma << 1$, making the Weibel modes stable for the realistic beam parameters. The suppression of the Weibel-type modes as shown by Rafighi et al. [6], left the electrostatic oblique modes as the relevant instability recovering the essential physics of the blazar-induced pair beams interactions with the IGM plasma [7]. It was demonstrated by Miniati and Elyiv [8], that the electrostatic modes are expected to develop in the kinetic regime. In the next section, we will introduce the relevant electrostatic instability in the kinetic regime.

3.4 The Beam-Plasma Electrostatic Instability

As we mentioned in Sect. 3.3, the electrostatic approximation is valid for the blazar-induced pair beam for which the electrostatic modes grow far more quickly than do the electromagnetic modes [7, 9]. A comparison of the Weibel growth rate for blazar-induced pair beams using a cold-beam distribution [10] and a Waterbag distribution [6] shows that the Weibel instability is suppressed for a realistic blazar-induced pair beam. Therefore, we will proceed with the electrostatic modes only in this section.

Assuming that the background plasma is cold and that the magnetic field is sufficiently weak such as $\omega_B \ll \omega_p$, where ω_B is the electron cyclotron frequency. Then, linearizing the Vlasov-Maxwell equations of the beam-plasma system for electrostatic waves leads to the following dispersion relation [11]

$$1 - \frac{\omega_p^2}{\omega^2} - \frac{4\pi n_b e^2}{k^2} \int d^3\mathbf{p} \frac{\mathbf{k} \cdot \frac{\partial f_b(\mathbf{p})}{\partial \mathbf{p}}}{\mathbf{k} \cdot \mathbf{v} - \omega} = 0, \tag{3.28}$$

where $f_b(\mathbf{p}) = f_b(\mathbf{p}, \mathbf{x})/n_b$ is the normalized momentum distribution function of the beam with n_b being the total number density of the beam. The wave vector is chosen as $\boldsymbol{k} = (k_\perp, 0, k_{||})$, and the beam propagates along the z axis with cylindrical symmetry.

There are two regimes that the instability can be characterized by, mainly, the reactive and the kinetic. The reactive regime is valid for a beam with negligible velocity spread or monochromatic beam $\Delta \mathbf{v} \approx 0$. In this case, the instability is hydrodynamic in nature and all the particles contribute to the growth rate, maximizing the growth rate along the longitudinal component of the wavevector to the beam direction. The peak reactive growth rate is given by [12]

$$\omega_{i,\mathrm{r}} = \frac{\sqrt{3}}{2^{4/3}} \omega_p \left(\frac{n_b}{\gamma n_e}\right)^{1/3} \left(\left(\frac{k_\perp}{k}\right)^2 + \frac{1}{\gamma^2}\left(\frac{k_{||}}{k}\right)^2\right)^{1/3}, \tag{3.29}$$

where γ is the beam Lorentz factor, and the parallel wave number is fixed at the resonance in this case, $k_{||} = \omega_p/c$.

On the other hand, if the beam velocity spread is significant, then the parallel electrostatic instability develops in the kinetic regime. In this case, only certain resonant particles participate in the growth of the unstable waves, and thus the growth rate is smaller than the reactive one. The kinetic instability is applicable if the velocity spread projection on the wave vector of the unstable waves is larger than the maximum growth rate of the reactive instability [7]

$$|\boldsymbol{k} \cdot \boldsymbol{\Delta v}| >> \omega_{i,\mathrm{r}}. \tag{3.30}$$

For a relativistic beam, the perpendicular velocity spread is $\Delta v_\perp \approx \frac{c}{\gamma}$, and the parallel velocity spread is $\Delta v_{||} \gtrsim \frac{c}{\gamma^2}$, resulting from the Lorentz boost of the beam from the center-of-momentum frame to the laboratory frame [8]. Given that for the realistic blazar-induced pair beams, the Lorentz factor is around $\gamma \sim 10^4 - 10^8$, and the beam to the background density ratio is around $\alpha = 10^{-17} - 10^{-12}$, the condition 3.30 is satisfied for essentially all oblique waves and marginally for the quasi-parallel waves, meaning that we should consider the kinetic regime for the realistic blazar-induced beams in the IGM and not the reactive one (cold-limit).

For the kinetic approach, the linear growth rate is evaluated by the Taylor expansion of the dispersion relation Eq. 3.28 for a small imaginary part of the frequency $\omega = \omega_R + i\omega_i$, and using the Cauchy theorem for the pole of the integrand in the dispersion relation 3.28. That yields the growth rate

$$\omega_i(\mathbf{k}) = \omega_p \frac{2\pi^2 e^2}{k^2} \int d^3\mathbf{p} \left(\mathbf{k} \cdot \frac{\partial f_b(\mathbf{p})}{\partial \mathbf{p}} \right) \delta(\omega_p - \mathbf{k} \cdot \mathbf{v}), \tag{3.31}$$

and $\omega_R = \omega_p$ for the real part of the frequency.

We note that the linear growth rate of the unstable waves with a wave vector **k** is found by the integral over the normalized momentum distribution function of the pair beam, $f_b(\mathbf{p})$, involving the resonance condition for the Cherenkov interaction between beam particle with velocity **v** and an electrostatic wave with a wave vector **k** and the plasma frequency ω_p

$$\omega_p - \mathbf{k} \cdot \mathbf{v} = 0. \tag{3.32}$$

Vafin et al. [13] demonstrated that the exact growth rate depends on the beam distribution function including both the beam angular and momentum profiles. Their results show that for a cold beam with no angular spread, the growth rate dominant modes are perpendicular to the beam, and when considering a realistic beam with finite angular spread, the growth rate is maximized at wave vectors quasi-parallel to the beam direction. The growth rate of the electrostatic waves (Eq. 3.31) is complicated to evaluate analytically for the realistic beam profile and it is often evaluated numerically as we will see in Chap. 4.

The linear growth rate is only applicable in the linear phase of the instability development. In this phase, the unstable plasma wave amplitude grows exponentially, $|\delta\mathbf{E}| \propto e^{\omega_i t}$, until the wave energy density is sufficient for the non-linear processes to interplay. In the following sections, we will introduce two of the main non-linear processes that influence the development of the blazar-induced beam electrostatic instability. The first one is the wave-particle non-linear scattering so-called non-linear Landau damping and the second one is the wave-wave interaction known as the modulation instability.

3.5 Non-linear Landau Damping

The overall non-linear evolution of the unstable waves generated by the blazar-induced pair beam is still unclear in the literature. Those non-linear processes are difficult to model including the whole processes; therefore, often they are studied by focusing on a certain process and neglecting others.

One of the relevant non-linear processes for the blazar-induced TeV beams is the non-linear Landau damping. This process describes the scattering of the resonant unstable plasma waves on the background plasma particles in the non-linear regime [14–16]. This process results in an energy and momentum exchange between the waves and the particles where energy is gradually transferred from the unstable plasma waves to the background particles. Resulting in the damping of the unstable wave amplitudes at the resonant wave vectors and shipping their energy to regions of smaller wavenumber.

The relevant process for the TeV pair beam is the scattering of the electrostatic waves on the background ions. Once the non-linear Landau regime is established, a thermal background ion with velocity, u_i, interacts with two electrostatic waves with frequencies $\omega(\mathbf{k})$ and $\omega(\mathbf{k}')$ through the Cherenkov interaction under the following condition

$$\left(\omega(\mathbf{k}) - \omega(\mathbf{k}')\right) - \left(\mathbf{k} - \mathbf{k}'\right) \cdot \mathbf{u}_i = 0. \tag{3.33}$$

Assuming a Maxwellian distribution of the background plasma, the non-linear landau damping rate of a wave with wave vector $\mathbf{k}$ due to the scattering into waves with the wave vectors $\mathbf{k}'$ is given by the following integral [15, 17]

$$\begin{aligned}\omega_{\rm NL}(\mathbf{k}) =& \frac{3\sqrt{2\pi}}{64} \int d^3\mathbf{k}' \frac{W(\mathbf{k}')}{n_e m_e u_i} \left(\frac{\mathbf{k}\mathbf{k}'}{kk'}\right)^2 \frac{k'^2 - k^2}{|\mathbf{k} - \mathbf{k}'|} \\ & \times \exp -\frac{1}{2}\left(\frac{3u_e^2}{2\omega_p u_i} \frac{k'^2 - k^2}{|\mathbf{k} - \mathbf{k}'|}\right)^2,\end{aligned} \tag{3.34}$$

where $u_{e,i} = \sqrt{T_{\rm IGM}/m_{e,i}}$ is the thermal velocity of the IGM electrons/ions, and $W(\mathbf{k}')$ is the spectral energy density of the electrostatic waves. Here, we assume that the IGM electrons and ions have the same temperature $T_{\rm IGM} = T_e = T_i$.

The electrostatic spectral energy density evolution is governed by the following non-linear wave kinetic equation incorporating the non-linear rate given by Eq. 3.34

$$\frac{dW(\mathbf{k})}{dt} = 2(\omega_i(\mathbf{k}) + \omega_{LL}(\mathbf{k}) + \omega_{NL}(\mathbf{k}) + \omega_c)W(\mathbf{k}), \tag{3.35}$$

where $\omega_i(\mathbf{k})$ is the linear electrostatic growth rate given by Eq. 3.31, and ω_{LL} is linear Landau damping rate given by Eq. 3.22 and can be rewritten as

$$\omega_{LL}(\mathbf{k}) = -\omega_p \sqrt{\frac{\pi}{8}} \left(\frac{\omega_p}{ku_e}\right)^3 \exp -\frac{1}{2}\left(\frac{\omega_p}{ku_e}\right)^2. \tag{3.36}$$

ω_c is the collisional damping rate [18]

$$\omega_c(k) = -\omega_p \frac{g}{6\pi^{\frac{3}{2}}} \frac{1}{(1 + 3k^2\lambda_D^2)^3}. \tag{3.37}$$

Here $g = (n_e\lambda_D^3)^{-1}$ is the plasma parameter, λ_D is Debye length, $n_e = 10^{-7}\text{cm}^{-3}$ is the density number of the intergalactic background electrons and $T_e = 10^4 K$ is the intergalactic medium temperature.

The non-linear Landau damping has been studied for the first time in the context of the blazar-induced pair beam by Miniati and Elyiv [8]. They used a simplified 1D model assuming the case of $k' \ll k$ where the waves are removed out of the resonance in a single scattering. Under this condition Eq. 3.34 reduces to

$$\omega_{\mathrm{NL,1D}} = \omega_p \frac{W_{\mathrm{nr}}}{n_e T_e} \frac{u_e^2}{u_i c}, \tag{3.38}$$

where W_{nr} is the total energy density of the electrostatic waves at the non-resonant wave numbers $k \ll \omega_p/c$. They also included the dissipation of the non-resonant waves by the Coulomb collisional damping, yielding the following equation for the evolution of the non-resonant waves

$$\frac{\partial W_{\mathrm{nr}}}{\partial t} = 2(\omega_{\mathrm{NL,1D}} W_{\mathrm{r}} + \omega_c), \tag{3.39}$$

where W_{r} is the total energy density of the electrostatic at resonant wave numbers $k \sim \omega_p/c$, which evolves as

$$\frac{\partial W_{\mathrm{r}}}{\partial t} = 2(\omega_{i,\mathrm{max}} W_{\mathrm{r}} - \omega_{\mathrm{NL,1D}} W_{\mathrm{r}}), \tag{3.40}$$

where $\omega_{i,\mathrm{max}}$ is the maximum linear growth rate of the electrostatic instability. Miniati and Elyiv [8] solved the coupled system of Eqs. 3.39 and 3.40 by assuming a steady state saturation of both equations, finding an energy loss time of the beam due to the instability that is much larger than the inverse Compton scattering energy loss time.

Afterwards, Chang et al. [19] solved numerical the kinetic wave evolution equation including the non-linear Landau damping and a simplified oblique growth rate. Their results indicated the opposite conclusion of Miniati and Elyiv [8], mainly that the non-linear Landau is not sufficient at suppressing the growth rate of the oblique instability. They concluded that the instability dominates the cooling of the pair beam before the inverse Compton cooling up to redshifts $z \approx 1$ and that it is less effective at higher redshifts.

Later, Vafin et al. [20] studied the impact of non-linear Landau damping including the electrostatic linear growth rate of the realistic blazar-induced pair beams. Where the linear growth rate decreases as the wavenumber increases, they used a simplified 2D model to solve numerically the kinetic evolution equation of the electric-field spectrum including the non-linear Landau.

Vafin et al. [20] found that the non-linear Landau damping transfers the entire energy of the unstable waves to smaller wave numbers that are not in resonance with the beam, concluding that the non-linear Landau damping prevents the unstable waves from taking a significant fraction of the beam energy during the inverse Compton cooling time. Vafin et al. [20] model was a 2D in the wavenumber space, where they considered the plane $k_{||} = \omega_p/c$ where $k_{||}$ is the wavevector component parallel with the beam propagation axis. The two dimensions they left are the perpendicular wave number and the azimuthal angle $(k_\perp, \varphi)$. The linear growth rate was limited to modes with $ck_\perp/\omega_p > 1$, whereas modes at $ck_\perp/\omega_p < 1$ were assumed to be stable.

3.5.1 3D Model of Non-linear Landau Damping

In the model of Vafin et al. [20], the parallel component of the wavevector was fixed to the resonance one, $k_{||} = \omega_p/c$, simplifying the problem to 2D in cylindrical coordinate $(k_\perp, \varphi, \omega_p/c)$. Here, we upgrade this model the full 3D space of the wave vector in the cylindrical coordinate $(k_\perp, \varphi, k_{||})$, where $k_\perp$ is the perpendicular component to the beam axis, φ is the polar angle of $k_\perp$ and $k_{||}$ is defined as the parallel component to the beam axis that defines the cylindrical symmetry axis.

Because of the azimuthal symmetry of the beam, then without losing the generality we can fix the wave vector $\mathbf{k}$ in Eq. 3.34 to $(\mathbf{k} = k_\perp \hat{y} + k_{||}\hat{z})$ and integrate over the full cylindrical coordinate of the scattered waves $\mathbf{k}'$ $(\mathbf{k}' = \sin\varphi' k'_\perp \hat{x} + \cos\varphi' k'_\perp \hat{y} + k'_{||}\hat{z})$, yielding the following 3D model of the non-linear landau damping rate

$$\omega_{\mathrm{NL}}(k_\perp, k_{||}) = \frac{3(2\pi)^{1/2}}{64 n_e m_e u_i} \int dk'_{||} \int dk'_\perp\, k'_\perp W(k'_\perp, k'_{||}) \mathrm{I}(k_\perp, k_{||}, k'_\perp, k'_{||}), \tag{3.41}$$

where

$$\begin{aligned} \mathrm{I}(k_\perp, k_{||}, k'_\perp, k'_{||}) = \int_0^\pi d\varphi' \frac{\left(k_\perp k'_\perp \cos\varphi' + k_{||}k'_{||}\right)^2}{\left(k_\perp^2 + k_{||}^2\right)\left(k'^2_\perp + k'^2_{||}\right)} F(k_\perp, k_{||}, k'_\perp, k'_{||}) \\ \times \exp -\frac{1}{2}\left(\frac{3u_e^2}{2\omega_p u_i} F(k_\perp, k_{||}, k'_\perp, k'_{||})\right)^2, \end{aligned} \tag{3.42}$$

and

$$F(k_\perp, k_{||}, k'_\perp, k'_{||}) = \frac{k'^2_\perp + k'^2_{||} - k_\perp^2 - k_{||}^2}{\sqrt{(k_\perp - k'_\perp \cos\varphi')^2 + (k_{||} - k'_{||})^2 + (k'_\perp \sin\varphi')^2}}. \tag{3.43}$$

The unstable modes are located in an extremely narrow range near the parallel wave number $k_{||} = \frac{\omega_p}{c}$. Therefore, we performed a nonuniform grinding for $ck_{||}\omega_p$ in the range 10^{-5}–1.5 with an exponential function $k_{||,i} = \exp -5\left(\frac{i-i_0}{i_0}\right)^n$, where the number of grid points is $i_{\max} = i_0\left(1 + \left(\frac{\ln 1.5}{5}\right)^{1/n}\right)$, $n = 7$ and $i_0 = 50$. We used logarithmic grinding for $ck_\perp/\omega_p$ for the range 10^{-5} up to 10^2 with 100 modes. The maximum linear growth rate we used here was of the order $\omega_{i,\max}^{-1} \approx 5 \times 10^6$ seconds and the IGM number density was $n_e = 10^{-7}$ cm^{-3}.

We calculated numerically Eq. 3.35 with the 3D form of the non-linear landau damping in Eq. 3.41. The initial condition for Eq. 3.35 is set to the thermal wave energy of the intergalactic medium $W_{\mathbf{k}}(t = 0) = k_B T_{\mathrm{IGM}} = 1$ eV per wavenumber mode $\mathbf{k}$. We also forced the solution $W_{\mathbf{k}}(t)$ to stay above the initial level $W_{\mathbf{k}}(t = 0)$ by the condition $W_{\mathbf{k}}(t = 0) \leq W_{\mathbf{k}}(t)$. The total electric field energy density is computed as

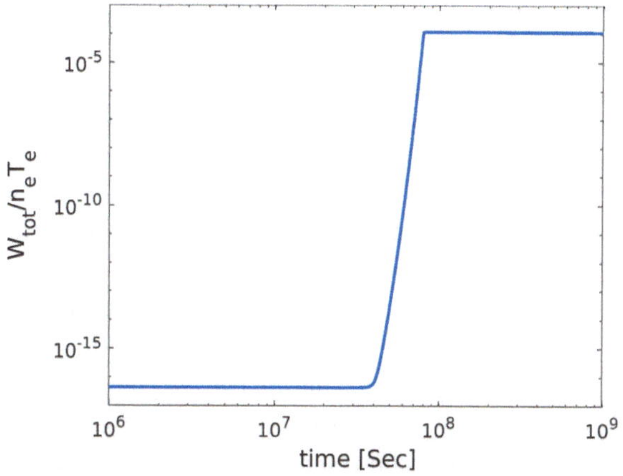

Fig. 3.2 The total electric field energy density normalized to the IGM energy density during the time evolution of Eqs. 3.35 and 3.41

$$W_{\mathrm{tot}} = 2\pi \int dk_{\parallel} \int dk_{\perp}\, k_{\perp} W(k_{\perp}, k_{\parallel}). \tag{3.44}$$

Figure 3.2 show the time evolution of the total electric field energy density during the evolution of the wave spectrum under the non-linear landau 3D model. During times much smaller than 10^8 seconds, the linear growth rate dominates and the unstable waves grow exponentially in the resonance region $ck_{\parallel}/\omega_p \approx 1$. In this regime the total electric field energy density increases as shown in Fig. 3.2. Afterwards, when the non-linear Landau damping starts operating, the waves are efficiently scattered to non-resonant smaller wave numbers. Once this non-linear regime starts, the spectral energy density of the resonant modes drops to the IGM background thermal fluctuations suppressing any efficient energy loss of the beam as it was shown in [20].

After the efficient shipping of the energy from the resonant to the non resonant modes, the waves of the non-resonant modes keep scattering to very small wave numbers under the impact of the non-linear landau scattering. However, the accumulation of these non-resonant waves can generate other non-linear processes like modulational instability. So far, we have only introduced and dealt with the non-linear Landau damping. In the next section, we will introduce another non-linear process called the modulation instability.

3.6 Modulation Instability

In the previous section, we introduced the non-linear landau damping resulting from the unstable electrostatic waves scattering on the background ions. The result of this process is the transfer of the energy of the unstable waves from the resonant modes to the non-resonant ones with small wave numbers. In this section, we look at modulation instability which is another non-linear process that could impact both the resonant and the non-resonant waves.

The modulation instability develops as the non-linear ponderomotive force of the unstable electrostatic waves produces cavitons (density cavities) [10, 21]. It results in the transformation of the energy from the beam resonant waves to the waves with shorter wavelengths. This modulation instability becomes relevant if the wave energy density of the resonant waves, W_{tot}, exceeds the threshold [19, 21]

$$\frac{W_{\text{tot}}}{n_e T_e} > 3(k\lambda_D)^2. \tag{3.45}$$

We see in Fig. 3.2 that the total electric field energy density exceeded the threshold for the modulation instability (where $3(k\lambda_D)^2 \approx 5 \times 10^{-6}$). This applies to both the resonant and the non-resonant waves. Where the non-resonant modes accumulate energy from the resonant when the non-linear Landau scattering efficiently operates. Therefore, including the modulation instability along with the non-linear Landau damping is necessary to understand the non-linear evolution of the instability.

The exact non-linear equations describing the modulation instability, i.e. equations 1-4 in Papadopoulos [22], are complex and often solved numerically. However, we can get an insight into the physics of the modulation instability by using the following simplified linearized damping rates for the resonant waves with total energy density, W_r, [23]

$$\omega_M(W_r) = \omega_p \frac{1}{4} \frac{W_r}{n_e T_e}, \tag{3.46}$$

for $3(k\lambda_{\text{D}})^2 < \frac{W_r}{n_e T_e} < \frac{m_e}{m_p}$, and

$$\omega_M(W_r) = \omega_p \sqrt{\frac{1}{3} \frac{m_e}{m_p} \frac{W_r}{n_e T_e}}, \tag{3.47}$$

for $\frac{W_r}{n_e T_e} > \max[3(k\lambda_{\text{D}})^2, \frac{m_e}{m_p}]$, where m_e and m_p are the electron and proton masses.

In the rest of this section, we combine this simplified description of the modulation instability with the simplified description of the non-linear Landau damping given by Eqs. 3.39 and 3.40 of [8].

3.6.1 *Waves Energy Density Evolution with Modulation Instability and Non-linear Landau Damping*

Since the resonant and non-resonant wave energy density during the non-linear Landau evolution exceeded the threshold for the modulation instability, we included in a simplified manner the modulation instability damping rate for the resonant and non-resonant modes. The evolution of the total energy density of the modes resonant with the beam can be described by the following equation

$$\frac{\partial W_r}{\partial t} = 2\omega_{i,\max} W_r - 2\omega_{NL}(W_{nr}) W_r - 2\omega_M(W_r) W_{mr}, \tag{3.48}$$

where the first term on the right-hand side is growth due to the beam resonant electrostatic instability, the second one is a non-linear Landau damping due to the non-resonant waves, W_{nr}, and the third term is also a damping rate due to the modulation. Here, we assume that the modulation shifts the energy to the modes called, W_{mr}. The non-resonant waves, W_{nr}, grow due to the non-linear Landau damping of the resonant waves and damping by the modulation to the waves noted, W_{mnr},

$$\frac{\partial W_{nr}}{\partial t} = 2\omega_{NL}(W_{nr}) W_r - 2\omega_M(W_{nr}) W_{mnr}. \tag{3.49}$$

Finally, the modes, W_{mr} and W_{mnr}, grow solely due to the modulation instability

$$\frac{\partial W_{mr}}{\partial t} = 2\omega_M(W_r) W_{mr}, \tag{3.50}$$

$$\frac{\partial W_{mnr}}{\partial t} = 2\omega_M(W_{nr}) W_{mnr}. \tag{3.51}$$

The non-linear damping rate, ω_{NL}, is given by Eq. 3.38 and the modulation damping rate, ω_M, is given by Eqs. 3.46 and 3.47. We have solved the system of equations (3.48–3.51) numerically with the initial condition where all wave mode regions are at a very low background level, $W(t = 0) = 10^{-20} n_e T_e$.

The result is shown in Fig. 3.3. We see that the resonance modes, W_r, grow exponentially until the non-linear Landau and the modulation instability operate. Shortly after the non-linear processes operate, the energy density of the resonance modes, W_r, saturates at the modulation threshold 3.45. This saturation is a result of balance on the three terms on the right-hand side of the Eq. 3.48. The non-resonant non-linear Landau modes, W_{mr}, also saturate but at a lower energy density than the resonance modes, W_r due to the balance between the non-linear Landau and the modulation terms on the right- hand side of the Eq. 3.49. In the end, the modes, W_{mr} and W_{mnr}, continue to grow in time feeding on the beam's kinetic energy. In this scenario, the energy of the TeV pair beams is efficiently transformed into an IGM heating by the instability.

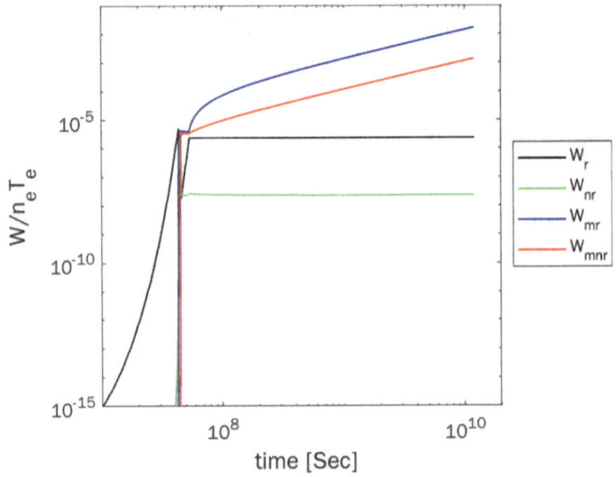

Fig. 3.3 The numerical solution of the system of Eqs. (3.48–3.51) describing the time evolution of the total energy density of the waves at different modes. The unstable resonant modes, W_r, are excited by the beam instability growth and damped by the non-linear Landau to the non-resonant modes, W_{nr} and the modulation instability to the modes, W_{mr}. The non-resonant modes, W_{nr}, are also damped by the modulation of the modes, W_{mnr}

References

1. Landau L (1946) On the vibrations of the electronic plasma. Zhurnal eksperimentalnoi i teoreticheskoi fiziki 16(7):574–586
2. L. A. Cottrill, A. B. Langdon, B. F. Lasinski, S. M. Lund, K. Molvig, M. Tabak, R. P. J. Town, and E. A. Williams. Kinetic and collisional effects on the linear evolution of fast ignition relevant beam instabilities. *Physics of Plasmas*, 15(8):082108, 08 2008. ISSN 1070-664X. https://doi.org/10.1063/1.2953816
3. A. Bret, L. Gremillet, and M. E. Dieckmann. Multidimensional electron beam-plasma instabilities in the relativistic regime. *Physics of Plasmas*, 17(12):120501, 12 2010a. ISSN 1070-664X. https://doi.org/10.1063/1.3514586
4. Bret A, Firpo M-C, Deutsch C (Oct2004) Collective electromagnetic modes for beam-plasma interaction in the whole k space. Phys. Rev. E 70:046401. URL https://link.aps.org/doi/10.1103/PhysRevE.70.046401
5. Weibel Erich S (Feb1959) Spontaneously growing transverse waves in a plasma due to an anisotropic velocity distribution. Phys. Rev. Lett. 2:83–84. URL https://link.aps.org/doi/10.1103/PhysRevLett.2.83
6. Rafighi I, Vafin S, Pohl M, Niemiec J (November 2017) Plasma effects on relativistic pair beams from TeV blazars: PIC simulations and analytical predictions. Astronomy & Astrophysics 607:A112. https://doi.org/10.1051/0004-6361/201731127
7. Chang Philip, Broderick Avery E, Pfrommer Christoph, Puchwein Ewald, Lamberts Astrid, Shalaby Mohamad, Vasil Geoffrey (2016) The Linear Instability of Dilute Ultrarelativistic $e^{\pm}$ Pair Beams. Astrophys. J. 833(1):118. https://doi.org/10.3847/1538-4357/833/1/118
8. Miniati Francesco, Elyiv Andrii (May2013) RELAXATION OF BLAZAR-INDUCED PAIR BEAMS IN COSMIC VOIDS. The Astrophysical Journal 770(1):54. https://doi.org/10.1088/0004-637x/770/1/54
9. Bret A, Gremillet L, Bénisti D (March 2010) Exact relativistic kinetic theory of the full unstable spectrum of an electron-beam-plasma system with Maxwell-Jüttner distribution functions. Phys. Rev. E 81(3):036402. https://doi.org/10.1103/PhysRevE.81.036402

10. Schlickeiser R, Ibscher D, Supsar M (Oct2012) PLASMA EFFECTS ON FAST PAIR BEAMS IN COSMIC VOIDS. The Astrophysical Journal 758(2):102. https://doi.org/10.1088/0004-637x/758/2/102
11. B.N. Breizman. Reviews of plasma physics. *Reviews of Plasma Physics*, 15, January 1990
12. Bret A, Gremillet L, Dieckmann ME (2010) Multidimensional electron beam-plasma instabilities in the relativistic regime. Physics of Plasmas 17(12):120501. https://doi.org/10.1063/1.3514586
13. S. Vafin, I. Rafighi, M. Pohl, and J. Niemiec. The electrostatic instability for realistic pair distributions in blazar/ebl cascades. *The Astrophysical Journal*, 857(1):43, Apr 2018. ISSN 1538-4357. https://doi.org/10.3847/1538-4357/aab552
14. V. N. Tsytovich and V. D. Shapiro. Nonlinear stabilization of plasma-beam instabilities. *Nuclear Fusion*, 5(3):228, sep 1965. https://doi.org/10.1088/0029-5515/5/3/006
15. Breĭzman BN, Ryutov DD, Chebotaev PZ (January 1972) Nonlinear Effects in the Interaction Between an Ultrarelativistic Electron Beam and a Plasma. Soviet Journal of Experimental and Theoretical Physics 35:741
16. Lesch H, Schlickeiser R (June1987) Stabilization and consequences of relativistic electron bumps in extragalactic radio sources. A&A 179:93–100
17. D. B. Melrose. *Instabilities in Space and Laboratory Plasmas*. 1986
18. Tigik SF, Yoon PH, Ziebell LF (Oct2019) Bremsstrahlung emission and collisional damping rate for langmuir waves. Plasma Physics and Controlled Fusion 61(12):125008. https://doi.org/10.1088/1361-6587/ab4aad
19. Chang Philip, Broderick Avery E, Pfrommer Christoph, Puchwein Ewald, Lamberts Astrid, Shalaby Mohamad (Dec2014) THE EFFECT OF NONLINEAR LANDAU DAMPING ON ULTRARELATIVISTIC BEAM PLASMA INSTABILITIES. The Astrophysical Journal 797(2):110. https://doi.org/10.1088/0004-637x/797/2/110
20. Vafin S, Deka PJ, Pohl M, Bohdan A (Feb2019) Revisit of nonlinear landau damping for electrostatic instability driven by blazar-induced pair beams. The Astrophysical Journal 873(1):10. https://doi.org/10.3847/1538-4357/ab017b
21. A. A. Galeev, R. Z. Sagdeev, Iu. S. Sigov, V. D. Shapiro, and V. I. Shevchenko. Nonlinear theory for the modulation instability of plasma waves. *Soviet Journal of Plasma Physics*, 1:5–10, January 1975
22. K. Papadopoulos. Nonlinear stabilization of beam plasma interactions by parametric effects. *The Physics of Fluids*, 18(12):1769–1777, 12 1975. ISSN 0031-9171. https://doi.org/10.1063/1.861096
23. H. P. Freund, I. Haber, P. Palmadesso, and K. Papadopoulos. Strongly turbulent stabilization of electron beam-plasma interactions. *The Physics of Fluids*, 23(3):518–527, 03 1980. ISSN 0031-9171. https://doi.org/10.1063/1.863000

Chapter 4
IGMF Impact on the Beam-Plasma Instability

The material in this chapter has been previously published in the publication Alawashra and Pohl [1].

In this chapter, we investigate the effect of weak IGMF with correlation lengths shorter than the pairs' inverse Compton cooling length on the plasma inability of the blazar-induced pair beam. The beam-plasma electrostatic instability operates best in the absence of an external magnetic field. Noting that magnetic deflection needs more than a femto-Gauss field amplitude to suppress the GeV-cascade, here we address the effect on the electrostatic instability that would be imposed by much weaker intergalactic magnetic fields. In particular, we investigate whether plasma instability still is the dominant energy-loss process and how strongly the cascade emission is suppressed by the instability.

In this study, we consider an IGMF with a small correlation length far below the energy-loss length of the pair beam, $\lambda_B \ll \lambda_e$, which deflects the electrons and positrons equivalently. Note that this condition implies that we assume the intergalactic magnetic fields to have no large-scale ($\gg$ kpc) or homogeneous component. We only consider the fluctuation component. Our IGMF model here is the same as that widely used in the analysis of deflection and time-delay limits [2–7]. Magnetic fields with strengths of $B_{\mathrm{IGM}} \ll 10^{-12}$ Gauss do not modify the linear dispersion relation of the beam-plasma instability obtained by the electrostatic approximation. However, those fields may impact the instability linear growth rate by their effect on the beam distribution function.

The focus lies on weaker field strength and on small correlation lengths. In such magnetic fields, the electrons and the positrons of the blazar-induced pair beam perform a random walk passing through many regions with different field orientations, resulting in an increased angular spread of the pair beam that scales with the mean field strength and the square root of the correlation length [8]. The case of large magnetic-field correlation lengths involves a net current in the beam and we do not consider it here.

M. S. A. Alawashra, *Plasma Instabilities of TeV Pair Beams Induced by Blazars*, Springer Theses, https://doi.org/10.1007/978-3-031-98568-3_4

We show in Sect. 4.1.2 that this widening of the beam significantly slows the electrostatic instability, which decreases the energy loss rate of the beam particles. At a certain limit in the parameter space ($B_{\rm IGM}$, $\lambda_{\rm B}$), driving the waves becomes less effective than inverse Compton scattering the CMB, and the GeV cascade emission can no longer be suppressed by the instability. For the plasma instability model in Vafin et al. [9], this limit is found to be around three orders of magnitude below the one that by magnetic deflection would impose a time delay of the cascade emission by 10 years [10].

In this study, we neglect the instability feedback on the beam as discussed in Perry and Lyubarsky [11]. We assume an efficient energy loss of the beam as discussed in Vafin et al. [9] and focus solely on the impact of the IGMF on this situation. The effect of the instability non-linear feedback on the beam is investigated and discussed in Chap. 5.

The structure of this chapter is as follows. In Sect. 4.1, we present the linear growth rate spectrum of the electrostatic instability of realistic pair beam distributions without and with weak intergalactic magnetic fields. In Sect. 4.2, we present the non-linear instability saturation of the unstable electrostatic waves. Finally, we demonstrate our results in Sect. 4.3 and conclude in Sect. 4.4.

4.1 Linear Growth Rate of the Electrostatic Instability

In this section, we present the linear growth rate of electrostatic waves for a realistic blazar-induced pair beam with finite angular spread (kinetic instability) moving in an unmagnetized intergalactic medium. Then we consider the magnetic fields in the intergalactic medium and find their impact on the beam distribution function and the implications for the growth rate of electrostatic waves.

As we mentioned in Chap. 3, the electrostatic approximation is valid for the blazar-induced pair beam for which the electrostatic modes grow far more quickly than do the electromagnetic modes [12, 13]. A comparison of the Weibel growth rate for blazar-induced pair beams using a cold-beam distribution Schlickeiser et al. [14] and a Waterbag distribution [15] shows that the Weibel instability is suppressed for a realistic blazar-induced pair beam. Therefore, we will proceed with the electrostatic approximation in our analysis.

For a relativistic electron beam ($\gamma_b >> 1$) with a small angular spread ($\Delta\theta << 1$ rad) travelling in a homogeneous background plasma with a number density n_e, the dispersion relation, Eq. 3.28, in the kinetic regime yields the following linear growth rate of electrostatic waves [16]

$$\begin{aligned}\omega_i(k) =& \pi\,\omega_p \frac{n_b}{n_e}\left(\frac{\omega_p}{kc}\right)^3 \int_{\theta_1}^{\theta_2} d\theta \\ &\times \frac{-2g(\theta)\sin\theta + (\cos\theta - \frac{kc}{\omega_p}\cos\theta')\frac{\partial g(\theta)}{\partial\theta}}{[(\cos\theta_1 - \cos\theta)(\cos\theta - \cos\theta_2)]^{1/2}},\end{aligned} \tag{4.1}$$

where

$$g(\theta) = m_e c \int_0^\infty dp\ p f_b(p,\theta), \tag{4.2}$$

and

$$\cos\theta_{1,2} = \frac{\omega_p}{kc}\left(\cos\theta' \pm \sin\theta' \sqrt{\left(\frac{kc}{\omega_p}\right)^2 - 1}\right), \tag{4.3}$$

where $k = \sqrt{k_\perp^2 + k_\parallel^2}$ is the module of the unstable electrostatic waves wave-number vector ($k_\perp$ and $k_\parallel$ are the perpendicular and the parallel components to the beam propagation direction respectively), θ' is the angle between the wave vector and the beam propagation direction, and θ is the angle between the particle momentum and the beam direction axis (z-axis). The beam is azimuthally symmetric around the propagation axis.

The beam distribution function, $f_b(p,\theta)$, of the beam is normalized as follows

$$2\pi \int_0^\infty dp\ p^2 \int_0^\pi d\theta\ \sin\theta f_b(p,\theta) = 1, \tag{4.4}$$

and can be factorized into parallel and perpendicular components

$$f_b(p,\theta) = f_{b,p}(p) f_{b,\theta}(p,\theta), \tag{4.5}$$

where for the parallel momentum distribution $f_{b,p}(p)$ we used equation 26 and equation 56 in Vafin et al. [9] that are obtained for a realistic pair beam at a distance of 50 Mpc from the blazar. The angular distribution, $f_{b,\theta}(p,\theta)$, depends on whether or not we have intergalactic magnetic fields.

4.1.1 Electrostatic Instability for a Pair Beam in Non-Magnetized Intergalactic Medium

In the case of a non-magnetized intergalactic medium, the angular spread of the beam is due to the angular energy spread only. In this case, the angular distribution function of the beam, $f_{b,\theta}(p,\theta)$, can be approximated by a Gaussian [17]

$$f_{b,\theta}(p,\theta) \approx \frac{1}{\pi \Delta\theta_s^2} \exp -\frac{\theta^2}{\Delta\theta_s^2}, \tag{4.6}$$

where the angular energy spread approximated as [18]

$$\Delta\theta_s \approx \frac{m_e c}{p}. \tag{4.7}$$

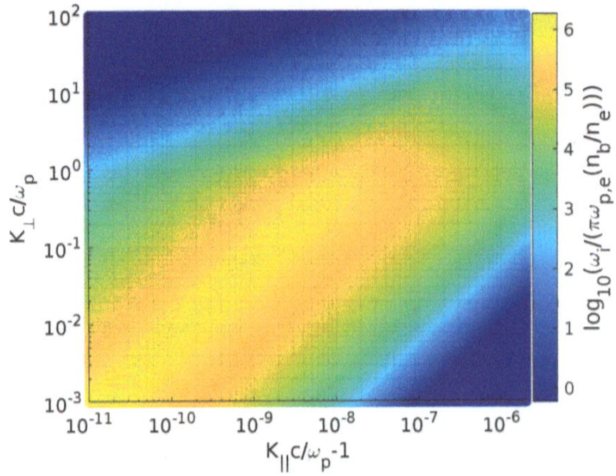

Fig. 4.1 Normalized electrostatic growth rate in the absence of an external magnetic field

Substituting Eq. 4.6 into the Eqs. 4.5 and 4.1 we found the numerical solution for the linear electrostatic growth rate as shown in Fig. 4.1, we see that most of the unstable modes are in the oblique and the parallel directions. We used the intergalactic background electrons unit density of $n_e = n_{e7}\ 10^{-7}\mathrm{cm}^{-3} = 10^{-7}\mathrm{cm}^{-3}$. For the fiducial pair beam parameters, the number density of the pair beam is $n_b = n_{b20}10^{-20}\mathrm{cm}^{-3} = 3 \times 10^{-22}\mathrm{cm}^{-3}$.

Vafin et al. [9] demonstrated that for a blazar with a redshift $z = 0.2$ those unstable waves drain the pair beam energy around a hundred times faster than does inverse Compton scattering on the CMB, taking into account the modulation instability as a damping process. The main uncertainties in that work are the assumptions on the spectrum and gamma-ray flux from the blazar and the approximation of the non-linear saturation level.

The growth rate is calculated for a pure electron beam moving in a background plasma of electrons and ions. Schlickeiser et al. [14] demonstrated that having separate distribution functions for electrons and for positrons yields the same growth rate as calculations that assume only an electron beam [18].

4.1.2 Electrostatic Instability for a Pair Beam with a Weak Intergalactic Magnetic Field

We address in this section the effects of weak intergalactic magnetic fields on electrostatic plasma instability. If the electrons' gyromagnetic frequency, $\omega_B = eB_{\mathrm{IGM}}/m_e$, is much smaller than their plasma frequency, $\omega_B \ll \omega_p$, then an external magnetic field doesn't change the electrostatic dispersion relation used to derive the linear growth rate [19]. The corresponding upper limit for the strength of the intergalactic magnetic field is $B_{\mathrm{IGM}} \lesssim 10^{-9}$ Gauss, where we used intergalactic medium number density of $n_e = 10^{-7}\ \mathrm{cm}^{-3}$.

The magnetic-field correlation lengths we consider, $\lambda_B \sim 10^3 - 10^{-5}$ pc, are much larger than the intergalactic plasma skin length, $\lambda_D \sim 5 \times 10^{-10}$ pc, meaning that even the variations of the IGMF have no direct impact on the beam plasma dispersion relation. However, the directional changes of the magnetic field affect the equilibrium beam distribution function, which in turn impacts the linear electrostatic growth rate (Eq. 4.1). In other words, the blazar-induced pair beam that triggers the instability travels through many correlation lengths in the IGMF.

Take the inverse Compton scattering length, $\lambda_{IC} \approx 75\,\mathrm{kpc}\left(10^7/\gamma_b\right)$, as an upper limit on the energy loss length of the beam particles, which gives around 188 kpc for a Lorentz factor of $\gamma = 4 \times 10^6$. This means that the pair beam distribution function carries the effects of the magnetic fields over a large number of directional changes, since most of the particles in the beam have travelled many correlation lengths at least, $\lambda_{IC} >> \lambda_B$. This propagation of the pair beam over many correlation lengths imposes an additional angular spread on its momentum distribution which in turn significantly affects the linear electrostatic growth rate.

Those fields lead to stochastic deflections of the electrons and positrons that diffusively widen the angular distribution function of the pair beam as shown in Appendix A.1. Adding in quadrature the energy angular spread $\Delta\theta_s$ (Eq. 4.7) and the magnetic widening $\Delta\theta_{IGMF}$ (Eq. A.8) gives the following distribution of the angular spread of the pair beam after travelling many correlation lengths in the IGM

$$f_{b,\theta}(\theta, p) = \frac{1}{\pi\,\Delta\theta^2}\exp-\Big(\frac{\theta}{\Delta\theta}\Big)^2, \quad 0 \le \theta \le \pi, \tag{4.8}$$

where

$$\Delta\theta = \frac{m_e c}{p}\sqrt{1 + \frac{2}{3}\lambda_B\lambda_{IC}\left(\frac{e B_{IGM}}{m_e c}\right)^2}. \tag{4.9}$$

Note that the result in Appendix A.1 for the magnetic deflection, $\Delta\theta_{IGMF}$, is consistent with the diffusion angle used in the IGMF deflection analyses, e.g. equation 31 in Neronov and Semikoz [20].

Finally, substituting Eq. 4.8 into the Eqs. 4.5 and 4.1 we numerically found the linear growth rate spectrum for a few values of the IGMF strength, B_{IGM}, and the correlation length, λ_B, and displayed it in Fig. 4.2. The main impact of the IGMF is a general reduction of the growth rate. Figure 4.3 shows the peak growth rate as a function of B_{IGM} and λ_B. To be noted from the figure is that specific values of the peak growth rate are found on a characteristic $B_{IGM} \propto \lambda_B^{-0.5}$. The reduction of the instability growth rate increases the energy loss time due to the plasma instability as we will see in the next section.

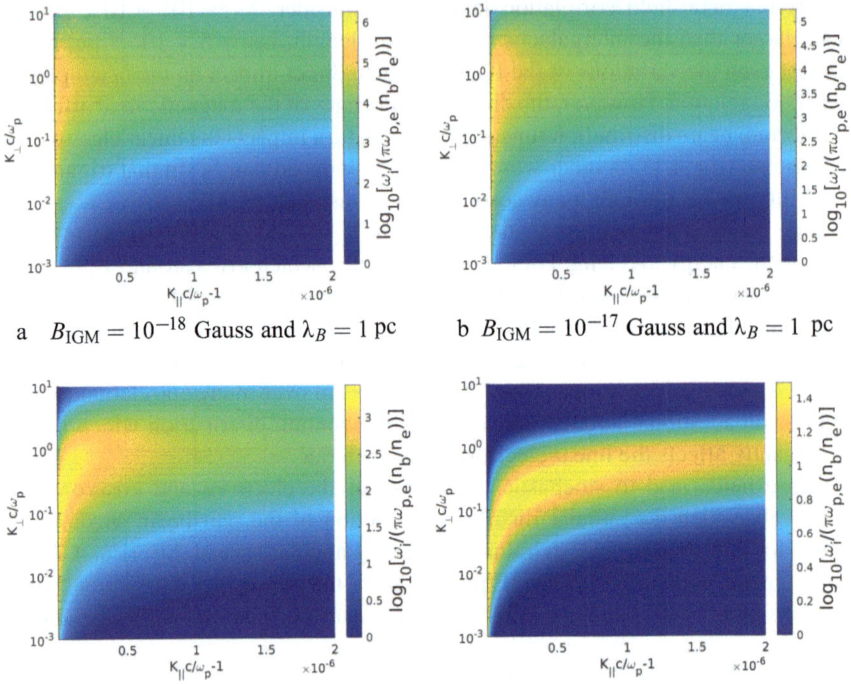

a $B_{\rm IGM} = 10^{-18}$ Gauss and $\lambda_B = 1$ pc

b $B_{\rm IGM} = 10^{-17}$ Gauss and $\lambda_B = 1$ pc

c $B_{\rm IGM} = 10^{-16}$ Gauss and $\lambda_B = 1$ pc.

d $B_{\rm IGM} = 10^{-15}$ Gauss and $\lambda_B = 1$ pc.

Fig. 4.2 The longtime of the normalized electrostatic growth rate ($\log_{10}(\omega_i/(\pi\omega_{p,e}(n_b/n_e)))$) for different intergalactic magnetic field strength values, we see that as the intergalactic magnetic fields strength increases the linear growth rate of the instability growth rate decreases

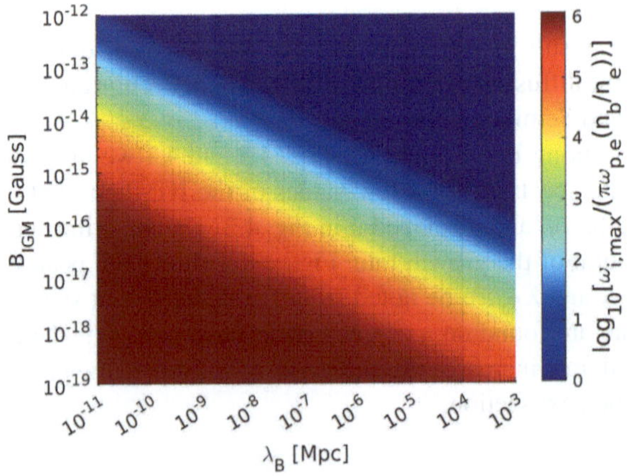

Fig. 4.3 The logarithm of the maximum normalized growth rate, $\log_{10}(\omega_{i,\max}/(\pi\omega_{p,e}(n_b/n_e)))$

4.2 Non-linear Instability Saturation

The unstable electrostatic waves grow exponentially with the linear growth rate, accumulating at the resonant parallel wave number $k_{||} \approx \omega_p/c$. Depending on their non-linear interactions, the waves could drain the kinetic energy of the pair beam and heat the IGM. The first type of non-linear interaction is the scattering of the electrostatic waves on the background plasma, known as non-linear Landau damping that we have discussed in Sect. 3.5. The second non-linear interaction is modulation instability which we introduced in Sect. 3.6.

Simulations of the evolution of the beam-plasma system are impossible right now for realistic parameters. However, there are various analytical estimates in the literature concerning the energy density that the waves reach in an equilibrium state [9, 14, 17, 18]. The inverse energy loss time of the pair beam due to the electrostatic instability is given by [9, 17]

$$\tau_{\text{loss}}^{-1} = 2\delta\omega_{i,\max}, \tag{4.10}$$

where $\omega_{i,\max}$ is the peak linear growth rate and $\delta = U_{\text{ES}}/U_{\text{beam}}$ is the normalized wave energy density at the equilibrium level. The reduction of the linear growth rate due to the IGMF translates into an increase in the energy loss time. At some limit, the beam-plasma instability becomes less relevant than the inverse Compton scattering. We will find this limit in the next section.

The wave intensity, δ, depends on the non-linear evolution of the electrostatic waves, which have different estimates. In the next section, we are going to include first the result given in Vafin et al. [9] and then discuss the implications of changing the value of the intensity of the waves to that found by Broderick et al. [18].

4.3 Results

We found the maximum linear growth rate of the unstable electrostatic 2D spectrum for each intergalactic magnetic field strength, B_{IGM}, and correlation length, λ_B, as shown in Fig. 4.3. Then we calculated the approximated energy loss time of the beam based on the maximum linear growth rate as in Eq. 4.10, using the intensity of the waves given in Vafin et al. [9], $\delta = 10^{-5}$. This time should be smaller than the inverse Compton scattering energy loss time, otherwise, the beam-plasma instability cannot suppress the secondary cascade. The energy loss time of the inverse Compton scattering is given by

$$\tau_{\text{IC}} = \left(7.7 \times 10^{13}\ \text{s}\right)\ (1+z)^{-4} \left(\frac{10^6}{\gamma}\right), \tag{4.11}$$

which at redshift $z = 0.2$ and for a pair-beam Lorentz factor $\gamma = 4 \times 10^6$ gives the following ratio for the beam-plasma instability loss time given in Vafin et al. [9]

$$\frac{\tau_{\rm loss}}{\tau_{\rm IC}} = 0.026, \tag{4.12}$$

if the intergalactic magnetic field is zero.

Using Eq. 4.10, we infer that a reduction by a factor 40 of the instability growth rate is sufficient to render it inefficient. The dependence of the growth rate on $B_{\rm IGM}$ and λ_B (see Fig. 4.3) can then be turned into a limit in the $B_{\rm IGM}$-λ_B parameter space, above which inverse Compton scattering provides the dominant energy loss of the pair beam. We show this limit in Fig. 4.4. It is at the same time an exclusion limit, because the then unavoidable inverse Compton emission is not seen, and so the cyan-shaded are in the figure is excluded for the IGMF. For weaker fields, the oblique instability may drain the beam energy sufficiently quickly, and for stronger fields, the time delay of the cascade emission causes substantial uncertainty in the interpretation of the Fermi-LAT data of GeV-scale cascade emission.

We see in Fig. 4.4 that the beam-plasma instability suppression limit (the purple line) is three orders of magnitude lower than the lower limit on the IGMF strength needed to impose a significant time delay of the cascade emission flux due to the magnetic deflection (the green line) [5, 10, 22]. We follow Ackermann et al. [10] in assuming a time period of 10 years as sufficient for the suppression of the cascade signal. The actual deflection angle would be well below one minute of arc.

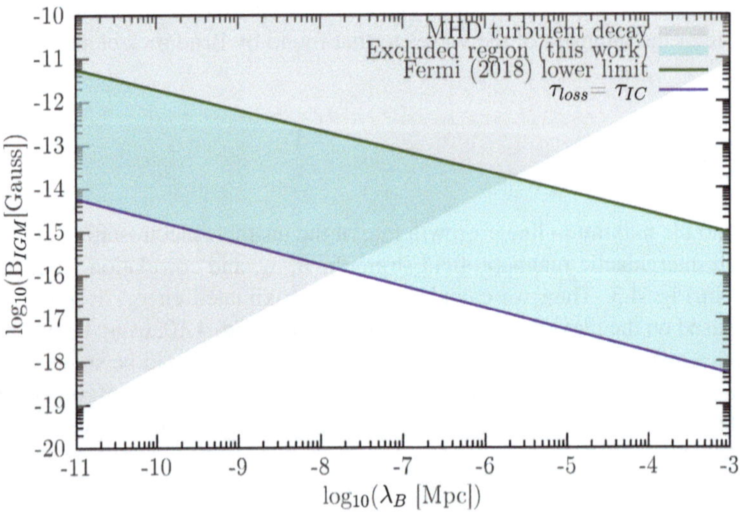

Fig. 4.4 The excluded region of the IGMF, for which neither magnetic deflection nor the oblique instability can explain the absence of cascade emission (Cyan). The grey region is the upper limit on the intergalactic magnetic field strength due to the MHD turbulent decay [8, 21]

Finally, to account for the uncertainty of the non-linear saturation level of the waves, we consider also the beam-plasma instability model presented in Broderick et al. [18] with $\delta = 0.2$ and check how the plasma instability suppression limit changes for a certain correlation length. For $\lambda_B = 10^{-11}$ Mpc, the instability limit would shift for Broderick et al. [18] to $B_{IGM} = 10^{-12.5}$ Gauss which is two orders of magnitude higher than that based on Vafin et al. [9] but still below the Fermi time delay lower limit. This shift applies also to all the instability suppression limit points in Fig. 4.4, since the correlation length and the intergalactic magnetic field strength determine together the angular spread Eq. 4.9 that plays the key role in determining the linear growth rate.

Although the non-linear saturation level had changed by four orders of magnitudes between the models of Vafin et al. [9] and Broderick et al. [18], the magnetic field limit had changed by only two orders of magnitude. This is due to the dependence of the energy loss time on the angular spread as $\tau_{\rm loss} \propto (\Delta\theta)^2$, which is a result of the maximum linear growth rate dependence on the angular spread as $\omega_{i,\rm max} \propto (\Delta\theta)^{-2}$ Vafin et al. [23], and the energy loss time relation with the linear growth rate as $\tau_{\rm loss} \propto \omega_{i,\rm max}^{-1}$ (Eq. 4.10).

For a generic beam-plasma instability model with plasma instability energy loss time $\tau_{\rm loss,0}$ in the absence of the IGMF, the energy loss time increases as $\tau_{\rm loss} \propto (\Delta\theta)^2$ when the angular spread increases with the IGMF strength and correlation length as in Eq. 4.9 reaching the inverse Compton scattering energy loss time at the following intergalactic magnetic field strength

$$\log\left(\frac{B_{\rm IGM,lim}}{\rm Gauss}\right) = -17.92 - \frac{1}{2}\log\left(\frac{\lambda_B}{\rm pc}\right) + \log\left(\sqrt{\frac{\rm Myr}{\tau_{\rm loss,0}}} - \sqrt{\frac{\rm Myr}{\tau_{\rm IC}}}\right). \tag{4.13}$$

Equation 4.13 provides the intergalactic magnetic fields strength that is sufficient to suppress a general plasma instability, with energy loss time $\tau_{\rm loss,0}$ in the absence of the IGMF, against the inverse Compton scattering of the blazar-induced pair beam on the CMB. For Vafin's model, the last logarithmic term on the right-hand side of Eq. 4.13 has a value very close to unity.

4.4 Summary

We investigated in this chapter the effects of tangled weak intergalactic magnetic fields with small correlation lengths on the electrostatic instability driven by blazar-induced pair beams. The weak fields increase the angular spread of the pair beam which decreases the linear growth rate of the electrostatic beam-plasma instability, which in turn reduces the associated energy loss rate.

In a certain region in the $B_{\rm IGM}$-λ_B parameter space, neither the beam-plasma instability nor the intergalactic magnetic field deflection can explain the absence of cascade emission in the spectra of some TeV blazars, and so this parameter space region can be excluded, unless there is a third mechanism that suppresses the GeV-band cascade.

Considering the beam-plasma instability model of Vafin et al. [9], we can exclude an IGMF strength within the three orders of magnitude below the limit above which magnetic deflection imposes a significant time delay of the cascade (ten years). Even for the non-linear evolution model of Broderick et al. [18], we can exclude a range a values that is one order of magnitude wide.

Although the parameter space below the beam-plasma instability suppression limit is not excluded by the cascade observations, part of this region (at $\lambda_{\rm B} \lesssim 10^{-8}$ Mpc and shaded in gray in Fig. 4.4 is constrained by MHD turbulent decay [8, 21]. In conclusion, the allowed region for the IGMF lies below 10^{-16} Gauss at $\lambda_{\rm B} \approx 10^{-8}$ Mpc, and the constraint on the IGMF strength is tighter than that for both larger correlation lengths (due to the instability suppression) and smaller correlation lengths (due to the MHD turbulent decay).

In this chapter, We have assumed an efficient energy loss of the beam as discussed in Vafin et al. [9] and focused on the impact of the IGMF on this instability energy loss. We have neglected the instability feedback on the beam. Perry and Lyubarsky [11] demonstrated that this instability feedback can influence the instability evolution significantly by broadening the beam. In the next chapter, we discuss this instability feedback in detail.

References

1. Mahmoud Alawashra and Martin Pohl. Suppression of the TeV Pair-beam–Plasma Instability by a Tangled Weak Intergalactic Magnetic Field. *The Astrophysical Journal*, 929(1):67, apr 2022. https://doi.org/10.3847/1538-4357/ac5a4b
2. A. Elyiv, A. Neronov, and D. V. Semikoz. Gamma-ray induced cascades and magnetic fields in the intergalactic medium. *Phys. Rev. D*, 80:023010, Jul 2009. https://doi.org/10.1103/PhysRevD.80.023010
3. A. Neronov and D. V. Semikoz. Sensitivity of γ-ray telescopes for detection of magnetic fields in the intergalactic medium. *Phys. Rev. D*, 80:123012, Dec 2009. https://doi.org/10.1103/PhysRevD.80.123012
4. Andrii Neronov and Ievgen Vovk. Evidence for Strong Extragalactic Magnetic Fields from Fermi Observations of TeV Blazars. *Science*, 328(5974):73, April 2010. https://doi.org/10.1126/science.1184192
5. A. M. Taylor, I. Vovk, and A. Neronov. Extragalactic magnetic fields constraints from simultaneous GeV-TeV observations of blazars. *A&A*, 529:A144, May 2011. https://doi.org/10.1051/0004-6361/201116441
6. Keitaro Takahashi, Masaki Mori, Kiyotomo Ichiki, and Susumu Inoue. LOWER BOUNDS ON INTERGALACTIC MAGNETIC FIELDS FROM SIMULTANEOUSLY OBSERVED GeV-TeV LIGHT CURVES OF THE BLAZAR mrk 501. *The Astrophysical Journal*, 744(1):L7, dec 2011. https://doi.org/10.1088/2041-8205/744/1/l7
7. Ie. Vovk, A. M. Taylor, D. Semikoz, and A. Neronov. FERMI /LAT OBSERVATIONS OF 1es 0229+200: IMPLICATIONS FOR EXTRAGALACTIC MAGNETIC FIELDS AND BACK-

GROUND LIGHT. *The Astrophysical Journal*, 747(1):L14, feb 2012. https://doi.org/10.1088/2041-8205/747/1/l14
8. Ruth Durrer and Andrii Neronov. Cosmological magnetic fields: their generation, evolution and observation. *The Astronomy and Astrophysics Review*, 21(1), Jun 2013. ISSN 1432-0754. https://doi.org/10.1007/s00159-013-0062-7
9. S. Vafin, I. Rafighi, M. Pohl, and J. Niemiec. The electrostatic instability for realistic pair distributions in blazar/ebl cascades. *The Astrophysical Journal*, 857(1):43, Apr 2018. ISSN 1538-4357. https://doi.org/10.3847/1538-4357/aab552
10. M. Ackermann, M. Ajello, L. Baldini, J. Ballet, G. Barbiellini, D. Bastieri, R. Bellazzini, E. Bissaldi, R. D. Blandford, E. D. Bloom, and et al. The search for spatial extension in high-latitude sources detected by the fermi large area telescope. *The Astrophysical Journal Supplement Series*, 237(2):32, Aug 2018. ISSN 1538-4365. https://doi.org/10.3847/1538-4365/aacdf7
11. Roy Perry and Yuri Lyubarsky. The role of resonant plasma instabilities in the evolution of blazar-induced pair beams. *MNRAS*, 503(2):2215–2228, May 2021. https://doi.org/10.1093/mnras/stab324
12. A. Bret, L. Gremillet, and D. Bénisti. Exact relativistic kinetic theory of the full unstable spectrum of an electron-beam-plasma system with Maxwell-Jüttner distribution functions. *Phys. Rev.*, 81(3):036402, March 010. https://doi.org/10.1103/PhysRevE.81.036402
13. Chang Philip, Broderick Avery E, Pfrommer Christoph, Puchwein Ewald, Lamberts Astrid, Shalaby Mohamad, Vasil Geoffrey (2016) The Linear Instability of Dilute Ultrarelativistic $e^{\pm}$ Pair Beams. Astrophys. J. 833(1):118. https://doi.org/10.3847/1538-4357/833/1/118
14. R. Schlickeiser, D. Ibscher, and M. Supsar. PLASMA EFFECTS ON FAST PAIR BEAMS IN COSMIC VOIDS. *The Astrophysical Journal*, 758(2):102, oct 2012. https://doi.org/10.1088/0004-637x/758/2/102
15. I. Rafighi, S. Vafin, M. Pohl, and J. Niemiec. Plasma effects on relativistic pair beams from TeV blazars: PIC simulations and analytical predictions. *Astronomy &Astrophysics*, 607:A112, November 2017. https://doi.org/10.1051/0004-6361/201731127
16. B.N. Breizman. Reviews of plasma physics. *Reviews of Plasma Physics*, 15, January 1990
17. Francesco Miniati and Andrii Elyiv. RELAXATION OF BLAZAR-INDUCED PAIR BEAMS IN COSMIC VOIDS. *The Astrophysical Journal*, 770(1):54, may 2013. https://doi.org/10.1088/0004-637x/770/1/54
18. Avery E. Broderick, Philip Chang, and Christoph Pfrommer. THE COSMOLOGICAL IMPACT OF LUMINOUS TeV BLAZARS. i. IMPLICATIONS OF PLASMA INSTABILITIES FOR THE INTERGALACTIC MAGNETIC FIELD AND EXTRAGALACTIC GAMMA-RAY BACKGROUND. *The Astrophysical Journal*, 752(1):22, may 2012. https://doi.org/10.1088/0004-637x/752/1/22
19. Ya B Fainberg. Interaction of charged particle beams with plasma. *Atomnaya Energ.*, 11, 10 1961. URL https://www.osti.gov/biblio/4815657
20. A. Neronov and D. V. Semikoz. Sensitivity of γ-ray telescopes for detection of magnetic fields in the intergalactic medium. *Physical Review D*, 80(12):123012, December 2009. https://doi.org/10.1103/PhysRevD.80.123012
21. Robi Banerjee and Karsten Jedamzik. Evolution of cosmic magnetic fields: From the very early universe, to recombination, to the present. *Phys. Rev. D*, 70:123003, Dec 2004. https://doi.org/10.1103/PhysRevD.70.123003
22. Justin D. Finke, Luis C. Reyes, Markos Georganopoulos, Kaeleigh Reynolds, Marco Ajello, Stephen J. Fegan, and Kevin McCann. CONSTRAINTS ON THE INTERGALACTIC MAGNETIC FIELD WITH GAMMA-RAY OBSERVATIONS OF BLAZARS. *The Astrophysical Journal*, 814(1):20, nov 2015. https://doi.org/10.1088/0004-637x/814/1/20
23. S. Vafin, P. J. Deka, M. Pohl, and A. Bohdan. Revisit of nonlinear landau damping for electrostatic instability driven by blazar-induced pair beams. *The Astrophysical Journal*, 873(1):10, feb 2019. https://doi.org/10.3847/1538-4357/ab017b

Chapter 5
Fokker-Planck Diffusion Simulation of the Instability Feedback

Part of the material in this chapter has been previously published in the article Alawashra and Pohl [1].

In this chapter, we study the instability feedback on the pair beam. This feedback was studied for the first time in the context of blazar-induced pair beam electrostatic instability by Perry and Lyubarsky [2]. The treatment in Perry and Lyubarsky [2] used a simplified one-dimensional beam distribution. In this chapter, we use instead a two-dimensional realistic beam distribution to explore the influence of the instability feedback on the beam. Specifically, we use the beam profile at a distance of 50 Mpc from the blazar found in Vafin et al. [3]. This treatment enables us to understand the feedback influence on the pairs with the relevant Lorentz factors for cascade emission in the GeV band.

The instability feedback is described as Fokker-Planck diffusion both in momentum and angular space. This treatment was simplified in the analysis by Perry and Lyubarsky [2], by evaluating only the initially dominant angular widening diffusion and neglecting the other effects involving the momentum diffusion and angular narrowing ($D_{\theta p}$). Here in Sect. 5.4.2, we check rigorously this assumption by using the 2D spectrum of the expanded beam under the dominant feedback to analyse the possible impact of the momentum diffusion on the beam energy and whether the beam narrowing is still negligible.

The blazar-induced pair beam-plasma instability significantly outpaces other factors that could change the beam profile, such as inverse Compton cooling and pair production. Whereas previous works have predominantly focused on assessing the instability's impact on a stationary beam profile, we incorporate the continuous production of TeV pairs into the transport equation of the beam, in addition to the instability diffusion in Sect. 5.4.4.

The structure of this chapter is as follows. In Sect. 5.1, we describe the numerical and analytical methods we used in calculating the time evolution of the beam-plasma system in the quasi-linear regime. In Sect. 5.2, we introduce the pair beam realistic 2D profile that we used in this study. In Sect. 5.3, we present the quasi-linear theory

M. S. A. Alawashra, *Plasma Instabilities of TeV Pair Beams Induced by Blazars*, Springer Theses, https://doi.org/10.1007/978-3-031-98568-3_5

of the beam-plasma system, we introduce the linear growth rate of the electrostatic instability, the time evolution of the electrostatic waves spectrum and the Fokker-Planck diffusion of the beam distribution. Finally, we demonstrate the numerical simulation and the results in Sect. 5.4 and conclude in Sect. 5.5.

5.1 Methodology

In this section, we present the numerical and analytic calculations we used to solve the time evolution of the beam-plasma system in the quasi-linear regime. We present the calculations of the linear growth rate of the electrostatic instability in Sect. 5.1.1. Then we introduce the diffusion coefficients calculations in Sect. 5.1.2. Finally, we demonstrate the numerical schemes for the Fokker-Planck diffusion equation of the instability feedback and the wave spectrum evolution equation in Sect. 5.1.3.

5.1.1 Modifications of Linear Growth Rate Calculations

In this section, we introduce our numerical and analytical calculations of the linear growth rate for a given beam distribution function. The linear growth rate is calculated for an unstable modes spectrum. The unstable wave modes are characterized by a wave number, k, and a wave vector angle with the beam propagation direction (z-axis), θ'. The wave vector azimuthal angle is fixed to zero, $\varphi' = 0$, without losing the generality due to the cylindrical symmetry of the beam. Furthermore, here we calculate the unstable modes in the Cartesian coordinates characterized by the perpendicular wave number, $k_\perp = k \sin\theta'$, and the parallel wave number, $k_{||} = k\cos\theta'$. The resonant unstable modes for relativistic blazar-induced pair beams are extremely narrow in the parallel wave number range from around $\frac{\omega_p}{c}\left(1 + 10^{-16}\right)$ to $\frac{\omega_p}{c}\left(1 + 10^{-6}\right)$. Therefore, often we will use the parallel wave number as, $\frac{ck_{||}}{\omega_p} - 1$.

In the previous Chap. 4, we were only interested in the maximum linear growth rate and its reduction with the IGMF. For that purpose, we used the calculations developed by Vafin and that have been used in the publications Vafin et al. [3, 4]. However, for the study of this chapter, we are also interested in more accurate information on the spectral shape of the linear growth rate and in using an arbitrary beam distribution as an input. Therefore, we improved several of the limitations of the calculations we previously adopted for the purposes of our study in this chapter.

The previous calculations assumed that the beam particle's speed is equal to the speed of light $v_b = c$ and hence there was no dependence on the beam Lorentz factor once evaluating the resonance angle boundaries. We found that the corrections of the beam particle speed become relevant once evaluating the linear growth rate for parallel unstable modes with the condition, $\left(\frac{ck_\perp}{\omega_p}\right)^2 + 2\left(\frac{ck_{||}}{\omega_p} - 1\right) \leq \frac{10^3}{2\gamma^2}$, where γ is the beam particle Lorentz factor (see Eq. B.5). We added this correction to both the linear growth rate integrated and the resonance boundaries. The previous code included the distribution function as an analytical form and therefore all the derivatives were performed analytically. For the calculations in this chapter, we used a numerical form of the distribution function with finite difference numerical derivatives up to the fourth order.

For an ultra-relativistic beam with beam particle speed $v_b \simeq c\left(1 - \frac{1}{2\gamma^2}\right)$, Eq. 3.31 yields

$$\omega_i(k_\perp, k_{||}) = \pi\omega_p \frac{n_b}{n_e}\left(\frac{\omega_p}{kc}\right)^3 m_e c \int_{p_{\min}}^{\infty} dp\; p\, I(p), \tag{5.1}$$

where

$$I(p) = \int_{\theta_1}^{\theta_2} d\theta \frac{-2f_b(p,\theta)\sin\theta + (\cos\theta - \frac{kv_b}{\omega_p}\cos\theta')\frac{\partial f_b(p,\theta)}{\partial\theta}}{[(\cos\theta_1 - \cos\theta)(\cos\theta - \cos\theta_2)]^{1/2}}, \tag{5.2}$$

and

$$\cos\theta_{1,2} = \frac{\omega_p}{kv_b}\left(\cos\theta' \pm \sin\theta'\sqrt{\left(\frac{kv_b}{\omega_p}\right)^2 - 1}\right), \tag{5.3}$$

θ' is the wave vector angle with the beam propagation direction (z-axis), θ is the beam's particle momentum angle with the beam propagation direction and $p = m_e c\gamma$ is the beam particle momentum. The lower bound on the beam momentum is given by the condition, $\left(\frac{kv_b}{\omega_p}\right)^2 \geq 1$, that yields

$$p_{\min} = \sqrt{\frac{1 + \left(\frac{ck_\perp}{\omega_p}\right)^2 + 2\left(\frac{ck_{||}}{\omega_p} - 1\right)}{\left(\frac{ck_\perp}{\omega_p}\right)^2 + 2\left(\frac{ck_{||}}{\omega_p} - 1\right)}}. \tag{5.4}$$

We calculated the linear growth rate in two different regimes in the parameter space ($\frac{ck_\perp}{\omega_p}$, $\frac{ck_{||}}{\omega_p} - 1$). The first one we call the quasi-parallel and oblique modes regime where $\left(\frac{ck_\perp}{\omega_p}\right)^2 > \left(\frac{ck_{||}}{\omega_p} - 1\right)$ and the second one is the parallel modes regime where $\left(\frac{ck_\perp}{\omega_p}\right)^2 \leq \left(\frac{ck_{||}}{\omega_p} - 1\right)$. In the first regime, the resonance beam angle region is very wide (see Fig. 5.1) and the calculations from the previous work of Vafin probably resolve the linear growth rate integrand. However, in the second regime, the resonance beam angle region is extremely narrow (see Fig. 5.1) and we developed a novel method for the calculations. First, we introduce the calculations in the quasi-parallel and oblique range and then we introduce the parallel one.

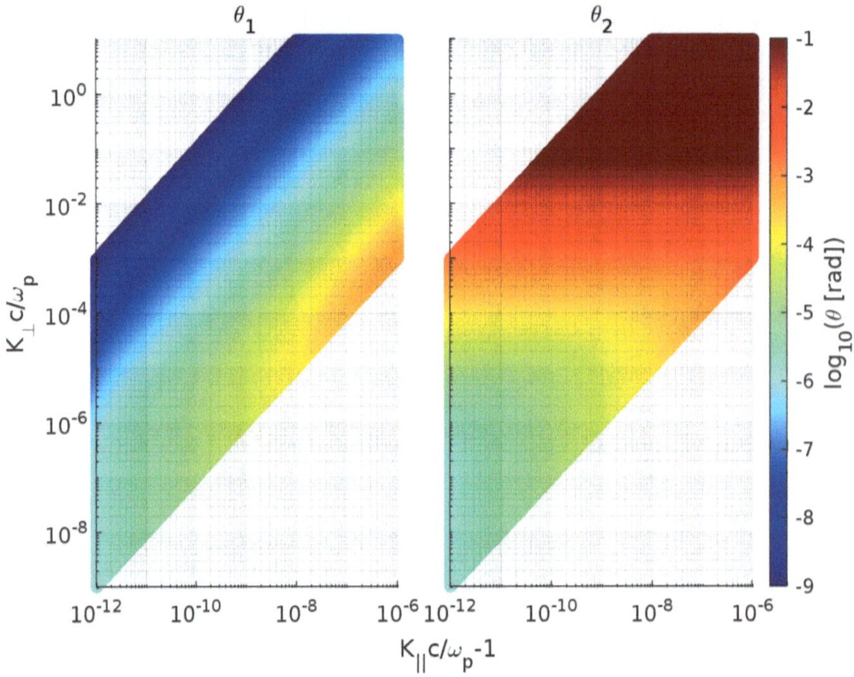

Fig. 5.1 The resonant beam angles boundaries (Eq. 5.3) for each wave mode ($k_\perp$, $k_{||}$). Here the beam Lorentz factor is fixed to $\gamma = 10^6$. Dark red areas are larger than 10^{-1} radian and blue dark are smaller than 10^{-9} radian

5.1.1.1 Quasi-Parallel and Oblique Modes Regime

In this regime, we consider the modes with the following condition, $\left(\frac{ck_\perp}{\omega_p}\right)^2 > \left(\frac{ck_{||}}{\omega_p} - 1\right)$, where the resonance region is very wide and the integral of Eq. 5.2 is well resolved by the beam angle variable, θ. In this case, we used the result from Vafin's calculations with several modifications. The original calculations assumed $v_b = c$, here we include the speed of the beam particles as $v_b \simeq c\left(1 - \frac{1}{2\gamma^2}\right)$. We rewrite Eq. 5.2 as

$$I(p) = \int_{\theta_1}^{\theta_2} d\theta \frac{F_\theta(p, \theta)}{[(\cos\theta_1 - \cos\theta)(\cos\theta - \cos\theta_2)]^{1/2}}, \tag{5.5}$$

where

$$F_\theta(p, \theta) = -2 f_b(p, \theta) \sin\theta + (\cos\theta - \frac{k v_b}{\omega_p} \cos\theta') \frac{\partial f_b(p, \theta)}{\partial \theta}. \tag{5.6}$$

To avoid the singularities at the boundaries during numerical integration of Eq. 5.5, the following method was proposed in the calculations by Vafin. The integral was split into two parts $[\theta_1, \theta_0]$ and $[\theta_0, \theta_2]$ where $\theta_0 = (\theta_1 + \theta_2)/2$. Then for the first integral interval, the relation, $[\cos\theta_1 - \cos\theta]^{-1/2} d\cos\theta = -2d[\cos\theta_1 - \cos\theta]^{1/2}$, were used applying partial integration. With the same thing applied to the second integral but using the other bracket the final result of Eq. 5.5 reads as

$$\begin{aligned} I(p) =& \frac{2F_\theta(p,\theta_0)}{\sin\theta_0}\sqrt{\frac{\cos\theta_1 - \cos\theta_0}{\cos\theta_0 - \cos\theta_2}} + \frac{2F_\theta(p,\theta_0)}{\sin\theta_0}\sqrt{\frac{\cos\theta_0 - \cos\theta_2}{\cos\theta_1 - \cos\theta_0}} \\ & - 2\int_{\theta_1}^{\theta_0} d\theta \frac{\sqrt{\cos\theta_1 - \cos\theta}}{\sin\theta\sqrt{\cos\theta - \cos\theta_2}} \left[\frac{dF_\theta(p,\theta)}{d\theta} - \frac{F_\theta(p,\theta)\left(2\cos\theta(\cos\theta - \cos\theta_2) - \sin^2\theta\right)}{2\sin\theta(\cos\theta - \cos\theta_2)}\right] \\ & + 2\int_{\theta_0}^{\theta_2} d\theta \frac{\sqrt{\cos\theta - \cos\theta_2}}{\sin\theta\sqrt{\cos\theta_1 - \cos\theta}} \left[\frac{dF_\theta(p,\theta)}{d\theta} - \frac{F_\theta(p,\theta)\left(2\cos\theta(\cos\theta_1 - \cos\theta) + \sin^2\theta\right)}{2\sin\theta(\cos\theta_1 - \cos\theta)}\right]. \end{aligned} \tag{5.7}$$

The original code, solving Eq. 5.7, was written using the analytical expressions of the first and second angular derivatives of the beam distribution function, $f_b(p, \theta)$. For the purpose of our calculations in this Chapter and since the initial beam's Gaussian angular profile evolves in time under the instability feedback we need to use numerical derivatives of the beam distribution allowing the beam profile to evolve in time. We used fourth-order sampled points finite difference numerical derivatives with a natural logarithm uniform grid of spacing, $\Delta \ln\theta$. The first-order angular derivative at the point, (p, θ_i), is given by

$$\left.\frac{\partial f_b(p,\theta)}{\partial\theta}\right|_{\theta_i} = \frac{f_{i-2} - 8f_{i-1} + 8f_{i+1} - f_{i+2}}{12\theta_i \Delta\ln\theta}, \tag{5.8}$$

where $f_i = f_b(p, \theta_i)$. Where the second-order angular derivative is given by

$$\begin{aligned} \left.\frac{\partial^2 f_b(p,\theta)}{\partial\theta^2}\right|_{\theta_i} = \frac{1}{12(\theta_i\Delta\ln\theta)^2}\Big[& -(\Delta\ln\theta + 1) f_{i-2} + (8\Delta\ln\theta + 16) f_{i-1} - 30 f_i \\ & + (-8\Delta\ln\theta + 16) f_{i+1} + (\Delta\ln\theta - 1) f_{i+2}\Big]. \end{aligned} \tag{5.9}$$

We used second-order sampled points finite difference numerical derivatives for the boundaries of our grid.

The resonance boundaries are defined by the Eq. 5.3. We found that in the regime where $\left(\frac{ck_\perp}{\omega_p}\right)^2 \gg 2\left(\frac{ck_\parallel}{\omega_p} - 1\right)$ and $\gamma^2\left(\frac{ck_\perp}{\omega_p}\right)^2 \gg 1$, the numerical values of $\theta_{1,2}$ as calculated in the old numerical code using the expression provided in Eq. 5.3 are susceptible to significant numerical inaccuracies. Expanding the square root to the leading orders is needed for robust calculations. We have employed this expansion for the expression of $\theta_{1,2}$, as detailed in the Appendix B.1 once we are in this regime. The final result for the resonance lower angle bound, θ_1, is given by Eq. B.10 where the resonance upper angle bound, θ_2, is given by Eq. B.11. We used Eqs. B.10 and B.11 to find the values of $\theta_{1,2}$, when $\left(\frac{ck_\perp}{\omega_p}\right)^2 \geq 200\left(\frac{ck_\parallel}{\omega_p} - 1\right)$ and $\gamma^2\left(\frac{ck_\perp}{\omega_p}\right)^2 \geq 100$, and Eq. B.7 otherwise.

We noted that the maximum linear growth rate value has been reduced by around a factor of 1.5 after implementing the new robust method of calculating $\theta_{1,2}$. We also found that there were significant numerical fluctuations in the linear growth rate map that disappeared completely after using the robust formula. This numerical correction is irrelevant to the previous study in Chap. 4, since we were interested in the reduction of the maximum linear growth rate by tangled intergalactic magnetic field over orders of magnitudes. A change in the linear growth rate maximum value by a factor of 1.5 would impact the suppression limits of the instability in Fig. 4.4 by around 1.5 for the IGMF correlation length, λ_B and by around $\sqrt{1.5}$ for the IGMF strength, $B_{\rm IGM}$. This shift is negligible compared to the uncertainties we had on the non-linear saturation of the instability.

5.1.1.2 Parallel Modes Regime

In the regime of $\left(\frac{ck_\perp}{\omega_p}\right)^2 \leq \left(\frac{ck_\parallel}{\omega_p} - 1\right)$, the resonant angles boundaries approach each other the value of $\theta_{1,2} \sim \sqrt{2}\left[\left(\frac{ck_\parallel}{\omega_p} - 1\right) - \frac{1}{2\gamma^2}\right]^{1/2}$ (see also Fig. 5.1), making the resonance region extremely narrow and impossible to resolve using the beam angle variable, θ. We therefore calculated the linear growth rate integral (Eq. 5.2) in this regime using a new variable, λ, defined as

$$\lambda = \frac{\theta^2}{2\left(\left(\frac{ck_\perp}{\omega_p}\right)^2 + \varepsilon_\parallel - \frac{1}{2\gamma^2}\right)}, \tag{5.10}$$

where the linear growth rate integrated numerically well resolved by linear grinding in λ. After the change of variable in the Appendix B.2, we arrived at the following result for $I(p)$ (Eq. 5.2) in terms of the variable, λ,

$$I(p) = \int_{\lambda_1}^{\lambda_2} d\lambda \frac{F_\lambda(p,\lambda)}{[(\lambda - \lambda_1)(\lambda_2 - \lambda)]^{1/2}}, \tag{5.11}$$

where

$$F_\lambda(p,\lambda) = -2f_b(p,\lambda) + \left(-\lambda + \frac{\frac{1}{2\gamma^2} - \varepsilon_\parallel}{\left(\left(\frac{ck_\perp}{\omega_p}\right)^2 + \varepsilon_\parallel - \frac{1}{2\gamma^2}\right)}\right)\frac{\partial f_b(p,\lambda)}{\partial\lambda}, \tag{5.12}$$

and the resonance boundaries for the variable, λ, are

$$\lambda_{1,2} = 1 \mp \frac{ck_\perp/\omega_p}{\left(\left(\frac{ck_\perp}{\omega_p}\right)^2 + \varepsilon_\| - \frac{1}{2\gamma^2}\right)} \sqrt{\left(\frac{ck_\perp}{\omega_p}\right)^2 + 2\varepsilon_\| - \frac{1}{\gamma^2}}. \tag{5.13}$$

We now apply the same mathematical trick we used for Eqs. 5.5 to 5.11 in order to avoid the singularities at the boundaries during the numerical integration. We split the integral into two parts $[\lambda_1, \lambda_0]$ and $[\lambda_0, \lambda_2]$ where $\lambda_0 = (\lambda_1 + \lambda_2)/2$. Then we integrate the first integral by parts using the relation, $[\lambda - \lambda_1]^{-1/2} d\lambda = 2d[\lambda - \lambda_1]^{1/2}$, and using the same thing for the second integral but using the other bracket we arrive to the final result of Eq. 5.11 as

$$\begin{aligned} I(p) =& 2F_\lambda(p, \lambda_0)\left[\sqrt{\frac{\lambda_0 - \lambda_1}{\lambda_2 - \lambda_0}} + \sqrt{\frac{\lambda_2 - \lambda_0}{\lambda_0 - \lambda_1}}\right] - 2\int_{\lambda_1}^{\lambda_0} d\lambda \sqrt{\frac{\lambda - \lambda_1}{\lambda_2 - \lambda}} \left[\frac{dF_\lambda(p,\lambda)}{d\lambda} + \frac{F_\lambda(p,\lambda)}{2(\lambda_2 - \lambda)}\right] \\ &+ 2\int_{\lambda_0}^{\lambda_2} d\lambda \sqrt{\frac{\lambda_2 - \lambda}{\lambda\ \ \lambda_1}} \left[\frac{dF_\lambda(p,\lambda)}{d\lambda} + \frac{F_\lambda(p,\lambda)}{2(\lambda - \lambda_2)}\right]. \end{aligned} \tag{5.14}$$

We have calculated the linear growth rate momentum integral in Eq. 5.1 using the Midpoint Riemann Sum with logarithmic grinding of the beam momentum, p. The resonance angles integral in Eq. 5.2 is solved in two cases: In the case of $\left(\frac{ck_\perp}{\omega_p}\right)^2 \geq 10\left(\frac{ck_\|}{\omega_p} - 1\right)$, we perform the integral calculations as in Eq. 5.7 using the Midpoint Riemann Sum with logarithmic grinding of the beam angle, θ. In the case $\left(\frac{ck_\perp}{\omega_p}\right)^2 < 10\left(\frac{ck_\|}{\omega_p} - 1\right)$, we use the result in Eq. 5.14 with linear grinding of the variable, λ, defined in Eq. 5.10. We found that the results of the two cases' calculations match each other at the transition line.

5.1.2 Diffusion Coefficients

The diffusion coefficients can be calculated using the following integral over the wave spectrum, $W(\mathbf{k})$, involving the resonance condition

$$D_{ij} = \pi e^2 \int d^3\mathbf{k} W(\mathbf{k}, t) \frac{k_i k_j}{k^2} \delta(\mathbf{k} \cdot \mathbf{v} - \omega_p), \tag{5.15}$$

where the unstable wave wavevector $\mathbf{k} = (k, \theta', \varphi')$ and the beam particles velocity $\mathbf{v}_b = (v_b \simeq c(1 - \frac{1}{2\gamma^2}), \theta, \varphi = 0)$ are both defined in the spherical coordinates with the beam propagation axis being the z-axis here. k_i is the projection of wave-vector $(\boldsymbol{k})$ to the spatial direction i. We use the azimuth symmetry of the wave spectrum and the beam to fix $\varphi = 0$ and integrate over φ' without losing the generality arriving at the following expression after a few steps shown in Appendix B.3

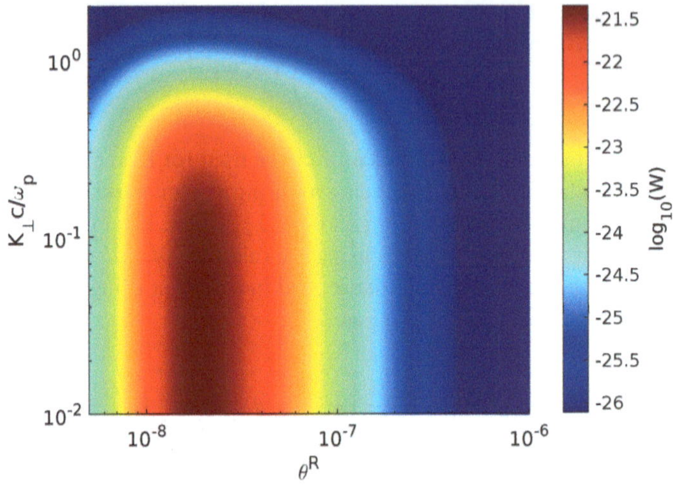

Fig. 5.2 A sample of the wave spectrum after a growth time of $7\omega_{i,\max}^{-1}$. We used here the linear growth rate resulting from the realistic initial beam profile as presented in Sects. 5.2 and 5.3.1. Here $W \equiv \frac{dW}{\frac{c}{\omega_p}\frac{ck_\perp}{\omega_p}d\frac{ck_\perp}{\omega_p}d\theta^R}$ in unites of eV cm^{-3}. We observe here that the wave spectrum is well-resolved numerically using the parameter $\theta^R = \left(\frac{ck_\parallel}{\omega_p} - 1\right) / \left(\frac{ck_\perp}{\omega_p}\right)$

$$\begin{Bmatrix} D_{pp} \\ D_{p\theta} \\ D_{\theta\theta} \end{Bmatrix} = \pi \frac{m_e\omega_p^2}{n_e} \int_{\omega_p/c}^{\infty} k^2 dk \int_{\cos\theta_1'}^{\cos\theta_2'} d\cos\theta' \frac{W(\mathbf{k})}{kv_b\sqrt{(\cos\theta' - \cos\theta_1')(\cos\theta_2' - \cos\theta')}} \begin{Bmatrix} 1 \\ \xi \\ \xi^2 \end{Bmatrix}, \tag{5.16}$$

where

$$\xi = \frac{\cos\theta\frac{\omega_p}{kv_b} - \cos\theta'}{\sin\theta}. \tag{5.17}$$

and the boundaries of $\cos\theta'$ are given by

$$\cos\theta_{1,2}' = \frac{\omega_p}{kv_b}\left[\cos\theta \pm \sin\theta\sqrt{\left(\frac{kv_b}{\omega_p}\right)^2 - 1}\right]. \tag{5.18}$$

We see in Eq. 5.16 that the integrand of the diffusion coefficients is proportional to the very narrow wave spectrum. We have resolved this narrow spectrum and the integrand numerically using a binning in the coordinates, $(k_\perp, \theta^R)$ where $\theta^R = \left(\frac{ck_\parallel}{\omega_p} - 1\right) / \left(\frac{ck_\perp}{\omega_p}\right)$ as shown in Fig. 5.2. Therefore, we had to transform the integral in Eq. 5.16 from the coordinates, (k, θ'), to the coordinates, $(k_\perp, \frac{ck_\parallel}{\omega_p} - 1)$, and then finally to the coordinates, $(k_\perp, \theta^R)$. The steps for those coordinates transformation are shown in the Appendix B.3 where the final expression of the diffusion coefficients is given by

$$\begin{Bmatrix} D_{pp} \\ D_{p\theta} \\ D_{\theta\theta} \end{Bmatrix} = \pi \frac{m_e \omega_p^2}{n_e c\theta} \int_{R(\theta,\gamma)} dk_\perp k_\perp \int_{R(\theta,\gamma)} d\theta^R$$

$$\times \frac{W(k_\perp, \theta^R)}{\sqrt{1 - \left(\frac{\theta^R}{\theta}\right)^2 + \frac{\theta^R}{ck_\perp/\omega_p}\left[1 + \left(\frac{1}{\gamma\theta}\right)^2\right] - \left(\frac{\omega_p}{ck_\perp}\right)^2\left[\frac{1}{2\gamma^2\theta} + \frac{\theta}{2}\right]^2}} \begin{Bmatrix} 1 \\ \xi \\ \xi^2 \end{Bmatrix}, \tag{5.19}$$

where

$$\xi = -\frac{1}{\sqrt{1 + 2\theta^R(ck_\perp/\omega_p) + (ck_\perp/\omega_p)^2(1 + \theta^{R^2})}}\left[\frac{\theta^R}{\theta}\frac{ck_\perp}{\omega_p} + \frac{\theta}{2} - \frac{1}{2\theta\gamma^2}\right], \tag{5.20}$$

and the resonance region $R(\theta, \gamma)$ is defined by the following condition in the parameter space, $(k_\perp, \theta^R)$, for each beam angle,θ, and Lorentz factor, γ

$$\left(\frac{ck_\perp}{\omega_p}\right)^2 \left(\theta^2 - \theta^{R^2}\right) + \frac{ck_\perp}{\omega_p}\theta^R\left[\theta^2 + \frac{1}{\gamma^2}\right] - \left[\frac{1}{2\gamma^2} + \frac{\theta^2}{2}\right]^2 \geq 0. \tag{5.21}$$

We have calculated the diffusion coefficients during the time evolution of the beam and waves using logarithmic grinding of the coordinates $(k_\perp, \theta^R)$ where the peak of the evolving wave spectrum is well resolved as well. We see in Eq. 5.19, that the dependence of the diffusion coefficients is very weak on the beam Lorentz factor, γ, for perpendicular wave numbers, $\frac{ck_\perp}{\omega_p} > 10^{-3}$. Since the unstable wave amplitude, W, is roughly constant for perpendicular wave numbers, $k_\perp$, less than $10^{-1}\frac{\omega_p}{c}$ and it is smaller for the parallel wave number modes, $k_\perp < 10^{-6}\frac{\omega_p}{c}$ (see Sect. 5.3.1), and since the diffusion coefficients integrands are proportional to $k_\perp \Delta k_\perp$ then we could safely neglect the contribution of modes with $k_\perp$ less than $10^{-3}\frac{\omega_p}{c}$. Therefore, we can also neglect the dependence of the diffusion coefficients on the beam Lorentz factor, γ, in this relevant range of $k_\perp$.

5.1.3 Beam Diffusion and Wave Evolution Equations

In this section, we introduce the numerical schemes we used to solve the time evolution of the beam and the wave spectrum during the Fokker-Planck diffusion of the beam-plasma system as presented in Sect. 5.4.1. We demonstrate in Sect. 5.4 that the $\theta\theta$ diffusion on the right-hand side of Eq. 5.40 is significantly larger than the other diffusion terms initially and therefore we will include only this term in this section

$$\frac{\partial f(p, \theta)}{\partial t} = \frac{1}{p^2\theta^2}\frac{\partial}{\partial \ln\theta}\left(D_{\theta\theta}\frac{\partial f(p, \theta)}{\partial \ln\theta}\right). \tag{5.22}$$

In Appendix C.1, we have performed the Von Neumann stability analysis of the diffusion equation (Eq. 5.22) for the FTCS (Forward Time Centered Space) numerical scheme. We found that the numerical stability of the FTCS scheme requires the following minimum time step

$$\Delta t \leq \min_i \left[p^2\theta_i^2 (\Delta \ln\theta)^2 \frac{(D^t_{\theta\theta i+\frac{1}{2}} + D^t_{\theta\theta i-\frac{1}{2}})}{(D'^2_{\theta\theta i+\frac{1}{2}} + D'^2_{\theta\theta i-\frac{1}{2}})} \right], \tag{5.23}$$

where θ_i is the beam angle with spatial index i computed using a logarithmic grid with spacing, $\Delta \ln\theta$. We define $D^t_{\theta\theta i+\frac{1}{2}} \equiv D_{\theta\theta}(\theta_{i+\frac{1}{2}}, t)$ and $D^{t+1}_{\theta\theta i-\frac{1}{2}} \equiv D_{\theta\theta}(\theta_{i-\frac{1}{2}}, t+\Delta t)$, where t is the initial time, Δt is the time step and $\theta_{i+\frac{1}{2}} = \exp{(\ln\theta)_{i+\frac{1}{2}}}$. We compute $(\ln\theta)_{i+\frac{1}{2}}$ as $(\ln\theta)_{i+\frac{1}{2}} = ((\ln\theta)_i + (\ln\theta)_{i+1})/2$.

We found that the numerical stability time step condition of the FTCS scheme (Eq. 5.23) gives time steps that are much shorter than the inverse linear growth rate times ω_i^{-1} making it very expensive to solve the diffusion equation using the FTCS numerical scheme. We also noted that the time step condition for the FTCS numerical (Eq. 5.23) is proportional to the inverse of the diffusion coefficients, $\Delta t \propto D^{-1}_{\theta\theta}(t)$, where the diffusion coefficients them self are proportional to the unstable wave amplitude (Eq. 5.15), $D_{\theta\theta} \propto W(t)$, that grows exponential during the linear phase of the instability evolution, $W(t) \propto \exp 2\omega_i t$, making the time step even shorter and shorter as the unstable waves grow. Therefore, we have adopted another numerical scheme to solve the beam diffusion equation, that is the **Crank–Nicolson** semi-implicit scheme [5].

We found in the Appendix C.2 that the numerical solution of Eq. 5.22 using the semi-implicit Crank-Nicolson scheme is unconditionally stable. However, the downside of this scheme is that it's much more difficult to solve than the FTCS scheme. In the case of the semi-implicit Crank-Nicolson scheme, one needs to solve a set of linear equations with a number of unknowns equal to the number of grid points.

Applying the semi-implicit Crank-Nicolson scheme with derivatives of the second order to Eq. 5.22, we get

$$\frac{f_i^{t+1} - f_i^t}{\Delta t} = \frac{1}{2p^2\theta_i^2(\Delta\ln\theta)^2} \left[D^t_{\theta\theta i+\frac{1}{2}}(f^t_{i+1} - f^t_i) - D^t_{\theta\theta i-\frac{1}{2}}(f^t_i - f^t_{i-1}) \right.$$
$$\left. + D^{t+1}_{\theta\theta i+\frac{1}{2}}(f^{t+1}_{i+1} - f^{t+1}_i) - D^{t+1}_{\theta\theta i-\frac{1}{2}}(f^{t+1}_i - f^{t+1}_{i-1}) \right], \tag{5.24}$$

where we define $f_i^t \equiv f(p, \theta_i, t)$ and $D^{t+1}_{\theta\theta i+\frac{1}{2}} \equiv D_{\theta\theta}(\theta_{i+\frac{1}{2}}, t+\Delta t)$ with t being the initial time, Δt is the time step and $\theta_{i+\frac{1}{2}} = \exp{(\ln\theta)_{i+\frac{1}{2}}}$. We compute $(\ln\theta)_{i+\frac{1}{2}}$ as $(\ln\theta)_{i+\frac{1}{2}} = ((\ln\theta)_i + (\ln\theta)_{i+1})/2$.

We can rewrite Eq. 5.24 as

$$-\xi D^{t+1}_{\theta\theta i+\frac{1}{2}} f^{t+1}_{i+1} + \left[1 + \xi\left(D^{t+1}_{\theta\theta i+\frac{1}{2}} + D^{t+1}_{\theta\theta i-\frac{1}{2}}\right)\right] f^{t+1}_i - \xi D^{t+1}_{\theta\theta i-\frac{1}{2}} f^{t+1}_{i-1} = C_i, \tag{5.25}$$

where

$$C_i = f^t_i + \xi\left[D^t_{\theta\theta i+\frac{1}{2}}(f^t_{i+1} - f^t_i) - D^t_{\theta\theta i-\frac{1}{2}}(f^t_i - f^t_{i-1})\right], \tag{5.26}$$

and $\xi = \frac{\Delta t}{2p^2\theta_i^2(\Delta \ln \theta)^2}$. The index i runs over the entire logarithmic grid of the beam angle, θ, giving as many linear equations in the form of Eq. 5.25 as many grid points, N_θ, we have. This results in a Tridiagonal and band-diagonal system of equations, that has nonzero elements only on the diagonal plus or minus one column. We solved this system of linear equations using the mathematical technique presented in Sect. 2.4 of Press [6].

We used the no-flux boundary condition for the distribution function at the minimum angle, $f^{t+1}_{-1} = f^{t+1}_0$, where the beam energy is conserved. As for the maximum beam angle, we used the condition, $f^{t+1}_{N_\theta+1} = 0$, where we stop the beam angle grid at a very small numerical value of 10^{-10} of the beam distribution peak value. We tested this with different values up to 10^{-15} of the beam distribution peak value.

For the wave spectrum evolution equation, we rewrite Eq. 5.35 as

$$\frac{\partial \ln W_{k,l}}{\partial t} = 2(\omega_{i\,k,l} - \omega_c), \tag{5.27}$$

where $W_{k,l} = \exp \ln W_{k,l}$ and we defined $W_{k,l} \equiv W(k_{\perp,k}, \theta_{R,l})$. We solved Eq. 5.27 using the FTCS (Forward Time Centered Space) scheme.

Lastly, we describe in detail how we combine the numerical solution of the beam diffusion equation (Eq. 5.22) with the solution unstable wave spectrum evolution equation (Eq. 5.27). We show a schematic diagram of the steps in Fig. 5.3. We start the simulation by calculating the instability linear growth rate, ω^t_i, at time t given the beam distribution function at time t, f^t, using Eq. 5.1. We then use this linear growth rate, ω^t_i, evolving the waves spectrum using Eq. 5.27 with the FTCS scheme from the time t, W^t, to the time $t+1$, W^{t+1}. Where $t+1$ represents the new time after a time step Δt.

We then use the wave spectrum at the time t, W^t, and the time $t+1$, W^{t+1}, to calculate the diffusion coefficients at the time t, $D^t_{\theta\theta}$, and at the time $t+1$, $D^{t+1}_{\theta\theta}$ respectively using Eq. 5.19. Afterwards, we use the diffusion coefficients at the current and later times ($D^t_{\theta\theta}$ and $D^{t+1}_{\theta\theta}$) along with the beam distribution at the current time, f^t, to find the beam distribution at the later time, f^{t+1}, using the Crank–Nicolson scheme of Eq. 5.22 as given by Eq. 5.25.

We then find the new linear growth rate, ω^{t+1}_i, using the updated beam distribution, f^{t+1}. Using this new growth rate we calculate the wave spectrum and the diffusion coefficients at the later time as before. We then updated the beam distribution for the new time step as before and so on as shown in Fig. 5.3. We used a dynamical time step of $\omega^{-1}_{i,\max}$ as the default time step with an upper limit set by the fastest rate of change of the beam distribution during the last simulation time step. We verified this by a convergence of the result by 0.1% using time steps that are 10 times smaller.

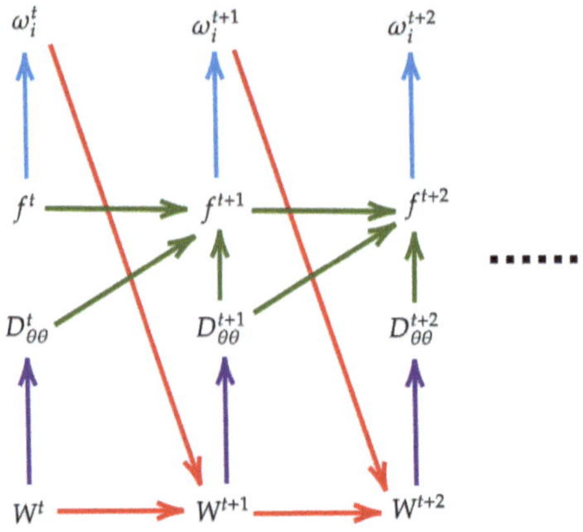

Fig. 5.3 Schematic diagram of the $\theta\theta$ Fokker-Planck feedback simulation steps. We present the simulation method in Sect. 5.1.3. The light blue arrows denote the resonance integration over the beam distribution, f, yielding the linear growth rate, ω_i, as given by Eq. 5.1. The purple arrows denote the resonance integration over the wave spectrum, W, yielding the diffusion coefficients, $D_{\theta\theta}$, as given by Eq. 5.19. The red arrows stand for the unstable wave spectrum, W, evolution according to Eq. 5.27 using the FTCS scheme. The green arrows represent the beam distribution function, f, evolution according to the diffusion Eq. 5.22 using the Crank–Nicolson scheme

For the wave spectrum, W, we use a logarithmic grid in the coordinates $(k_\perp, \theta^R)$ where $\theta^R = \left(\frac{ck_\parallel}{\omega_p} - 1\right) / \left(\frac{ck_\perp}{\omega_p}\right)$. We used 100 grid points for the perpendicular wave number, $k_\perp$, from $10^{-3}\ \frac{\omega_p}{c}$ to $10\ \frac{\omega_p}{c}$, we have verified a convergence of this by using 300 points. For the parameter, θ^R, we used 600 grid points for the interval 10^{-9} to 5×10^{-3} where we have tested this with 1500 grid points. For the beam distribution, f, we use a logarithmic grid in the coordinates (θ, γ) where γ is the beam particle Lorentz factor. We used 100 grid points for γ from 10^4 to 10^8 and verified a convergence of this with 300 grid points. Finally for the beam particle angle, θ, we used 600 grid points from 10^{-9} radian to 5×10^{-3} radians tested by using 1500 grid points.

5.2 Blazar-Induced Pair Beam Distribution

The pair-beam distribution function is the crucial quantity that determines the beam-plasma instability growth rate [3]. Thus, using the realistic spectrum of blazar-induced pair beams is essential for examining the influence of the beam-plasma instability on the beam and the GeV-scale cascade emission. In this study, we used the realistic beam distribution at a distance of 50 Mpc from the blazar, as reported in Vafin et al. [3]. Here, we introduce this beam spectrum and explain the ingredients used to find it.

The propagation of the beam distribution in the IGM is driven by two primary factors. The first one is the pair's production due to the interaction of the high-energy gamma rays with the EBL, along with their subsequent cooling processes. The second effect is the dispersion of the primary gamma-ray flux with the propagation distance, leading to an inverse proportionality of the beam density with the square of the distance from the blazar. These two fundamental mechanisms collectively shape the evolution of the pair-beam distribution along the propagation distance in the IGM.

The two effects have been combined in Vafin et al. [3], neglecting the IC cooling, to calculate the accumulated pair spectrum over the IC cooling mean-free path of pairs with Lorentz factor of 10^7 starting at the distance 50 Mpc from the blazar. The neglect of IC cooling is driven by the necessity to investigate beam-plasma instabilities that provide the dominant energy loss. They used an intrinsic power-law gamma-ray spectrum with a spectral index of 1.8 and a cut-off step function at the energy of 50 TeV.

We define the normalized beam momentum distribution, $f(p, \theta) = f(\mathbf{x}, \mathbf{p})/n_b$

$$\int d^3p \; f(p, \theta) = 1, \tag{5.28}$$

where n_b is the pair-beam density. The density at 50 Mpc from the blazar is estimated as $n_b = 3 \times 10^{-22}$ cm^{-3} [3]. Factorizing the distribution as

$$f(p, \theta) = \frac{d^3 f}{dp^3} = \frac{1}{2\pi m_e c \beta p^2} f_\gamma(\gamma) f_{\cos\theta}(\gamma, \theta), \tag{5.29}$$

where $f_{\cos\theta}(\gamma, \theta)$ is an angular differential part, $f_\gamma(\gamma)$ is a parallel momentum differential part, and β is the normalized speed. The angular part is approximated by a Gaussian [3, 7, 8]

$$f_{\cos\theta}(\gamma, \theta) = \frac{2}{\Delta\theta^2} \exp -\left(\frac{\theta}{\Delta\theta}\right)^2, \tag{5.30}$$

with the angular spread of $\Delta\theta = \frac{1}{\gamma}$. The parallel momentum part, $f_\gamma(\gamma)$, is given by Eq. D.1 in the Appendix D, where we replaced the sharp step-function cut-off used in Vafin et al. [3] by an exponential cut-off at the Lorentz factors higher than 6×10^6 using a part of a logarithmic Gaussian as shown in Fig. D.1.

The initial normalized realistic beam spectrum is shown in Fig. 5.4, and the main energy bulk of the pair is located at Lorentz factors of a few 10^6. We also see that the pairs are concentrated in a narrow band around the production angles of $\theta \sim \gamma^{-1}$. We will use this distribution to find the linear growth rate of the instability in Sect. 5.3.1 and as the initial condition for the Fokker-Planck simulation of the instability feedback in Sect. 5.4.

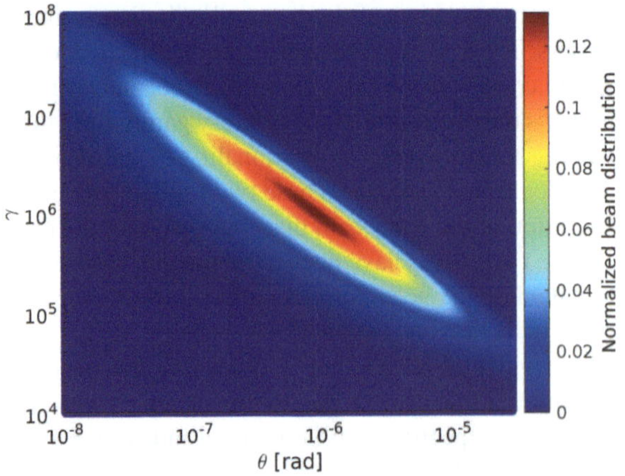

Fig. 5.4 The normalized initial beam distribution $2\pi p^3\theta^2 f(p,\theta)/n_b$ at distance 50 Mpc from the blazar [3]

5.3 Quasi-Linear Theory of the Beam-Plasma System

The beam-plasma instabilities manifest in both electrostatic and electromagnetic modes, including the two-stream instability ($\mathbf{k} \times \delta\mathbf{E} = 0$ where $\delta\mathbf{E}$ is the perturbed electric field), the transverse Weibel, and filamentation modes ($\mathbf{k} \cdot \delta\mathbf{E} = 0$) [9]. The electrostatic modes dominate the wave spectrum for the blazar-induced TeV beams, whereas Weibel-type modes are suppressed [10, 11] (see Sect. 3.3). Consequently, we consider only the electrostatic oblique modes, which is sufficient to recover the essential physics [12].

In Sect. 5.3.1, we present the linear growth rate of the electrostatic instability. In Sect. 5.3.2, we introduce the balance equation for waves. Lastly, in Sect. 5.3.3, we introduce the Fokker-Planck diffusion equation describing the instability feedback on the beam distribution.

5.3.1 *Electrostatic Linear Growth Rate*

In the kinetic regime that is applicable for blazar-induced pair beams [8], the linear growth rate of an unstable wave with a wave vector $\mathbf{k}$ can be found by [13]

$$\omega_i(\mathbf{k}) = \omega_p \frac{2\pi^2 e^2}{k^2} \int d^3\mathbf{p} \left(\mathbf{k} \cdot \frac{\partial f(\mathbf{p})}{\partial \mathbf{p}}\right) \delta(\omega_p - \mathbf{k} \cdot \mathbf{v}_b), \tag{5.31}$$

where $\omega_p = (4\pi n_e e^2/m_e)^{1/2}$ is the plasma frequency of the intergalactic background plasma with density n_e. Exploiting the cylindrical symmetry around the beam propagation axis (z-axis in our case), we fixed the wave vector of the electrostatic waves to $\mathbf{k} = (k_\perp, 0, k_{\|})$, where $k_\perp$ and $k_{\|}$ are the perpendicular and the parallel components to the beam propagation direction respectively. After integrating over the azimuthal angle of the beam we get the following

$$\omega_i(k_\perp, k_{||}) = \pi\omega_p \frac{n_b}{n_e}\left(\frac{\omega_p}{kc}\right)^3 \int_{p_{\min}}^{\infty} dp m_e c\, p \int_{\theta_1}^{\theta_2} d\theta \\ \times \frac{-2f(p,\theta)\sin\theta + (\cos\theta - \frac{kv_b}{\omega_p}\cos\theta')\frac{\partial f(p,\theta)}{\partial\theta}}{[(\cos\theta_1 - \cos\theta)(\cos\theta - \cos\theta_2)]^{1/2}}, \tag{5.32}$$

where the boundaries are given by

$$\cos\theta_{1,2} = \frac{\omega_p}{kv_b}\left(\cos\theta' \pm \sin\theta'\sqrt{\left(\frac{kv_b}{\omega_p}\right)^2 - 1}\right), \tag{5.33}$$

and

$$p_{\min} = \sqrt{\frac{1 + \left(\frac{ck_\perp}{\omega_p}\right)^2 + 2\left(\frac{ck_{||}}{\omega_p} - 1\right)}{\left(\frac{ck_\perp}{\omega_p}\right)^2 + 2\left(\frac{ck_{||}}{\omega_p} - 1\right)}}. \tag{5.34}$$

Here θ' is the wave vector angle with the beam propagation direction (z-axis), θ is the angle between momentum and the beam axis, and $v_b \simeq c(1 - \frac{1}{2\gamma^2})$ is the particle speed.

In Fig. 5.5, we present the linear growth rate (Eqs. 5.32 and 5.33), using the beam distribution introduced in Sect. 5.2, the density of IGM electron as $n_e = 10^{-7}(1+z)^3\mathrm{cm}^{-3}$, and a redshift $z = 0.15$. Previous treatments in the literature used the approximation ($v_b = c$) when evaluating Eq. 5.32 [2, 3, 8]. However, we found that in the regimes of wave numbers with the condition, $\left(\frac{ck_\perp}{\omega_p}\right)^2 + 2\left(\frac{ck_{||}}{\omega_p} - 1\right) \leq \frac{10^3}{2\gamma^2}$, the difference between the particle speed and the speed of light becomes relevant (see Eq. B.5). We have taken this difference into account in our calculations of Fig. 5.5 where the detailed numerical and analytical calculations are presented in Sect. 5.1.1.

In Fig. 5.5, we see that the growth rate is maximal and constant in the range of perpendicular wave numbers, $10^{-6} < ck_\perp/\omega_p < 1$, with a sharp drop at the oblique angles $ck_\perp/\omega_p \sim 1$, whereas for the parallel modes, $k_\perp c/\omega_p < 10^{-6}$, it is smaller by around a factor of 3. Note that the growth rate of the parallel modes is sensitive to the beam distribution, for a simplified monoenergetic Gaussian the growth of the parallel modes is larger than the quasi-parallel ones [2], where it's smaller for the distribution we used and for a Maxwell–Jüttner distribution [12].

The turnover of the linear growth rate spectral shape around the wave numbers of $\left(\frac{ck_\perp}{\omega_p}\right)^2 \sim \left(\frac{ck_{||}}{\omega_p} - 1\right)$ is due to the change in the corresponding resonant beam angles. In the regime of $\left(\frac{ck_\perp}{\omega_p}\right)^2 \gg \left(\frac{ck_{||}}{\omega_p} - 1\right)$, the resonant beam angles are constrained by

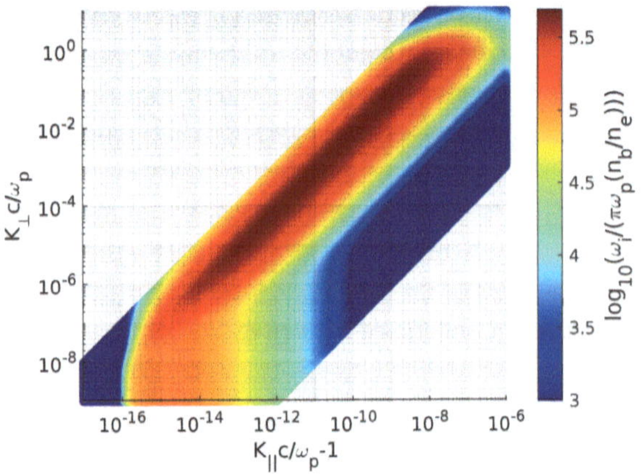

Fig. 5.5 Normalized linear growth rate using the realistic beam distribution function. White areas denote stable modes

the minimum angles of $\theta_1 \simeq \left(\frac{ck_{\parallel}}{\omega_p} - 1\right) / \left(\frac{ck_{\perp}}{\omega_p}\right)$ and large θ_2 (see Fig. 5.1). In the regime of $\left(\frac{ck_{\perp}}{\omega_p}\right)^2 \ll \left(\frac{ck_{\parallel}}{\omega_p} - 1\right)$, the resonant angles boundaries approach each other to a $ck_{\perp}/\omega_p$ independent value of $\theta_{1,2} \sim \sqrt{2}\left[\left(\frac{ck_{\parallel}}{\omega_p} - 1\right) - \frac{1}{2\gamma^2}\right]^{1/2}$.

The maximum growth rate, $\omega_{i,\max} \sim 6.7 \times 10^{-8}\ \mathrm{s}^{-1}$, is much faster than the IC cooling rate of the beam, $\tau_{\mathrm{IC}}^{-1}(\gamma) \approx \gamma \times 1.3 \times 10^{-20}(1+z)^4\ \mathrm{s}^{-1}$. However, the instability-induced energy-loss rate significantly depends on the nonlinear evolution of the instability [4, 8, 14, 15].

In this study we focus on the instability feedback, therefore we will consider only the linear regime of the instability and neglect the restrictions on the growth of the waves due to non-linear interactions. In the next section, we briefly introduce the linear evolution equation of unstable waves and levels of the unstable wave's energy density where the nonlinear processes become relevant.

5.3.2 *Evolution of the Wave Spectrum*

The quasi-linear evolution of the wave spectrum for homogeneous plasma is governed by the following equation

$$\frac{\partial W(\mathbf{k})}{\partial t} = 2(\omega_i(\mathbf{k}) + \omega_c(k))W(\mathbf{k}), \tag{5.35}$$

where $W(\mathbf{k})$ is the spectral energy density of the electric field oscillations, $\omega_i(\mathbf{k})$ is the linear growth rate as defined in Sect. 5.3.1, and ω_c is the collisional damping rate [16],

$$\omega_c(k) = -\omega_p \frac{g}{6\pi^{3/2}} \frac{1}{(1+3k^2\lambda_D^2)^3}. \tag{5.36}$$

Here $g = (n_e\lambda_D^3)^{-1}$ is the plasma parameter, $\lambda_D = 6.9\,\mathrm{cm}\sqrt{\frac{T_e/K}{n_e/\mathrm{cm}^{-3}}}$ is the Debye length, $n_e = 10^{-7}(1+z)^3\mathrm{cm}^{-3}$ is the density of IGM electrons, and $T_e = 10^4 K$ is their temperature. We start integrating Eq. 5.35 at the very low thermal fluctuations level.

The collisional damping rate given by Eq. 5.36 is approximately 20 times smaller than the approximation employed in other studies (i.e. Miniati and Elyiv [8]; Perry and Lyubarsky [4]; Vafin et al. [2]), which did not account for the microscopic wave-particle interactions. Those interactions were included under the generalized weak turbulence theory in Yoon et al. [17], deriving an accurate general kinetic formulation of the collisional damping rate of the electrostatic plasma waves, that was used in Tigik et al. [16] to find the collisional damping rate (Eq. 5.36).

The total electric field energy density is calculated by

$$W_{\mathrm{tot}} = 2\pi \int dk_\perp k_\perp \int dk_{||} W(k_\perp, k_{||}). \tag{5.37}$$

Accounting for the energy equipartition between kinetic electrostatic fluctuations, the energy loss rate of the beam due to the growth of the electrostatic waves at time t is given by [3]

$$\begin{aligned}\frac{dU_b}{dt}(t) =& -2\frac{dW_{\mathrm{tot}}}{dt}(t)\\ =& -8\pi \int dk_\perp k_\perp \int dk_{||} W(k_\perp, k_{||}, t)\omega_i(k_\perp, k_{||}, t),\end{aligned} \tag{5.38}$$

where the total beam energy density is defined as

$$U_b = 2\pi \int dp p^2 \int d\theta \sin\theta m_e c^2 \gamma f(p, \theta). \tag{5.39}$$

In the previous section, we found that the modes with the maximum growth, $10^{-6} < ck_\perp/\omega_p < 1$, grow at the same rate (see Fig. 5.5), maintaining a similar spectral amplitude. However, the energy density in those modes is proportional to their wave number volume element, $2\pi k_\perp \Delta k_\perp \Delta k_{||}$ (Eq. 5.37). Therefore, we can focus on the quasi-parallel and oblique modes, $10^{-3} < ck_\perp/\omega_p < 1$, since they dominate the energy density of the unstable mode spectrum. We can also neglect inhomogeneity of the background plasma since it is relevant only for the strictly parallel modes [2, 18].

The amplitude of the unstable modes grows exponentially until their wave intensity is high enough to trigger nonlinear processes. One of the main non-linear processes is the modulation instability that moves wave energy from resonant to non-resonant modes. This process operates when the total electric field energy density hits the threshold of around $(k\lambda_D)^2 n_e T_e$ [8], resulting in the saturation of the resonant unstable mode at around 10^{-3} of the total beam energy we consider here.

Another non-linear process is non-linear Landau damping, where the non-linear scattering of the unstable waves on the background plasma ions results in severe damping of the resonant modes. Non-linear Landau damping becomes effective when the total electric field energy density reaches around 10^{-2} of that of the beam [4, 15].

The impact of these non-linear interactions is still uncertain [4, 8, 15, 19]. The numerical noise in simulations (such as PIC) is too high, and the numerical growth rate is too small, for a reliable assessment, on account of the very small beam density. Upscaling of the beam density and downscaling the beam Lorentz factor is possible, but the results of those simulations are difficult to scale back to the realistic parameters [11, 20].

In this work, we focus on the nonlinear feedback of the instability on the beam and so we consider only the linear phase of the instability growth. We discuss in Sect. 5.4 that under the instability feedback on the beam, the total electric field energy density stays always below the non-linear thresholds for the beam density we consider. In the next section, we look at the Fokker-Planck diffusion equation that describes the feedback of the electrostatic waves on the beam during the quasilinear regime.

5.3.3 Fokker-Planck Diffusion Equation for the Pair Beam

The quasilinear regime is applicable when the total wave energy density is much smaller than that of the plasma. In this regime, the feedback of the electrostatic unstable waves on the beam is governed by the following Fokker-Planck diffusion equation [21]

$$\begin{aligned}\frac{\partial f(p,\theta)}{\partial t} =& \frac{1}{p^2\theta}\frac{\partial}{\partial\theta}\left(\theta D_{\theta\theta}\frac{\partial f}{\partial\theta}\right) + \frac{1}{p\theta}\frac{\partial}{\partial\theta}\left(\theta D_{\theta p}\frac{\partial f}{\partial p}\right)\\ &+ \frac{1}{p^2}\frac{\partial}{\partial p}\left(p D_{p\theta}\frac{\partial f}{\partial\theta}\right) + \frac{1}{p^2}\frac{\partial}{\partial p}\left(p^2 D_{pp}\frac{\partial f}{\partial p}\right),\end{aligned} \tag{5.40}$$

where the diffusion coefficients are defined by the following resonance integrals [22]

$$D_{ij}(\mathbf{p}) = \pi e^2 \int d^3\mathbf{k} W(\mathbf{k})\frac{k_i k_j}{k^2}\delta(\mathbf{k}\cdot\mathbf{v} - \omega_p), \tag{5.41}$$

where the electric charge, e, is given in cgs units. The pair-beam distribution function, f, is given in spherical coordinates (p, θ, φ), and so is the wave-vector $\mathbf{k}$ (k, θ', φ'). The angles θ and θ' are defined with respect to the beam propagation direction $(z-\text{axis})$. Due to the azimuthal symmetry of the pair-beam distribution function, we can set $\varphi = 0$ and integrate over φ', yielding (see Appendix B.3)

$$\begin{Bmatrix} D_{pp} \\ D_{p\theta} \\ D_{\theta\theta} \end{Bmatrix} = \frac{\pi m_e \omega_p^2}{n_e} \int_{\omega_p/c}^{\infty} k^2 dk \int_{\cos\theta_1'}^{\cos\theta_2'} d\cos\theta' \frac{W(\boldsymbol{k})}{k v_b} \times \frac{1}{\sqrt{(\cos\theta' - \cos\theta_1')(\cos\theta_2' - \cos\theta')}} \begin{Bmatrix} 1 \\ \xi \\ \xi^2 \end{Bmatrix}, \tag{5.42}$$

where

$$\xi = \frac{\cos\theta \frac{\omega_p}{k v_b} - \cos\theta'}{\sin\theta}, \tag{5.43}$$

and $v_b \simeq c(1 - \frac{1}{2\gamma^2})$ is the particle speed for Lorentz factor γ. The boundaries of the $\cos\theta'$ integration are fixed by the resonance condition

$$\cos\theta_{1,2}' = \frac{\omega_p}{k v_b} \left[\cos\theta \pm \sin\theta \sqrt{\left(\frac{k v_b}{\omega_p}\right)^2 - 1} \right]. \tag{5.44}$$

The integrands are largest at the peak of the wave spectrum, therefore a proper numerical resolution of the spectrum is necessary when calculating the diffusion coefficients. We have changed the integration variables in Sect. 5.1.2 arriving at the coordinates $(k_\perp, \theta^R)$ with $\theta^R = \left(\frac{ck_\parallel}{\omega_p} - 1\right) / \left(\frac{ck_\perp}{\omega_p}\right)$, for which the peak of the unstable modes is numerically well resolved and the diffusion coefficients are well defined by Eq. 5.19. We also found that the Lorentz factor, γ, dependence of the diffusion coefficients is negligible compared to the beam angle, θ.

In the next section, we describe the numerical setup of our study of the instability feedback and present our results.

5.4 Numerical Results

We have calculated the rate of change for every term on the right-hand side of the Fokker-Planck Eq. 5.40, using the diffusion coefficients of a wave spectrum generated after around 20 growth times, $20\omega_{i,\max}^{-1}$, of the growth rate presented in Sect. 5.3.1. We represent those rates by the following notion, $\tau_{ij}^{-1} = \frac{1}{f}\left(\frac{\partial f}{\partial t}\right)_{ij}$, where the indices i and j loop over θ and p with the correspond terms on the RHS of Eq. 5.40.

We found that the diffusion term $D_{\theta\theta}$ exceeds the other terms by orders of magnitude in the phase-space region containing the bulk of the beam particles. We see this clearly in Fig. 5.6 where we plotted the highest three rates for beam Lorentz factors, $\gamma = 10^6$ and 10^7. Evaluating the maximum rate across the entire parameter space for every term, we found the following ratio between the different terms: $\theta\theta : \theta p : p\theta : pp \approx 1 : 10^{-3} : 10^{-5} : 10^{-8}$.

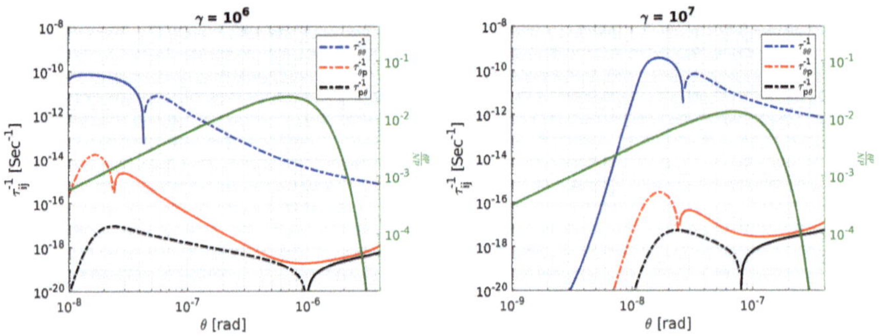

Fig. 5.6 The rate of change of the beam distribution for the largest three terms of the RHS of the Fokker-Planck Eq. 5.40 for a given wave spectrum generated after around 20 growth times, $20\omega_{i,\max}^{-1}$. We see here that the diffusion $\theta\theta$ significantly outpaces the diffusion θp and $p\theta$. Solid lines are negative rates, where particles get removed from those angles, and dash-doted lines are positive rates, where particles are increased at those angles. Note that this is only a snapshot of a given wave spectrum for the purpose of comparison of the different terms, in the actual feedback simulation the exponential wave growth is coupled time-dependently with the beam diffusion

Given this result, we initially neglect all the subdominant terms and in Sect. 5.4.1 consider only the diffusion term $D_{\theta\theta}$. We will check the validity of this approximation in Sect. 5.4.2 as we analyse the effect of the subdominant terms as the $\theta\theta$ diffusion modifies the beam. We also analyse the dependence of our results on the beam parameters in Sect. 5.4.3. Finally, we add the continuous pair production to our simulation setup in Sect. 5.4.4.

5.4.1 Simulation of the θθ Angular Diffusion Feedback

Having established that $D_{\theta\theta}$ initially dominates over the other diffusion terms by orders of magnitudes, we perform here a numerical simulation of instability feedback including only this term. We introduce the simulation setup in Sect. 5.4.1.1 and present the results in Sect. 5.4.1.2.

5.4.1.1 Simulation Setup

The numerical simulation of the beam-plasma system only includes the first term on the right-hand side of Eq. 5.40

$$\frac{\partial f(p,\theta)}{\partial t} = \frac{1}{p^2\theta}\frac{\partial}{\partial\theta}\left(\theta D_{\theta\theta}\frac{\partial f(p,\theta)}{\partial\theta}\right), \tag{5.45}$$

coupled time-dependently with the waves' spectral evolution Eq. 5.35. The linear growth rate of the instability (Eq. 5.32) and the diffusion coefficients (Eq. 5.42) involve integration over the beam distribution function and the wave spectrum, respectively.

We solve Eq. 5.45 using the Crank–Nicolson scheme along with the FTCS scheme for the wave equation, Eq. 5.35. We have demonstrated the simulation steps in Sect. 5.1.3. We used a dynamical time step of $\omega_{i,\max}^{-1}$ as the default time step with an upper limit set by the fastest rate of change of the distribution. We tested this by using time steps that are 10 times smaller. To properly resolve the narrow wave spectrum we use a logarithmic grid in the coordinates $(k_\perp, \theta^R)$ where $\theta^R = \left(\frac{ck_\parallel}{\omega_p} - 1\right) / \left(\frac{ck_\perp}{\omega_p}\right)$. We verified convergence in our grid resolution for both the wave spectrum and the beam distribution. The initial beam distribution is as described in Sect. 5.2, and the initial wave energy density corresponds to the fluctuation level [4].

5.4.1.2 Results

We found that the instability feedback severely increased the beam's angular spread. This broadening strongly depends on the Lorentz factor of the beam particles. In Fig. 5.7, we show the angular spread for different beam Lorentz factors. We see that particles with larger Lorentz factors get scattered earlier since those particles are in resonance with faster-growing wave modes, and so the scattering feedback affects them earlier.

The angular spreading of the beam immediately shifts the resonant wave numbers. In Fig. 5.8, we see the reduction of the growth rate for the parallel wave numbers during the simulation time for a fixed perpendicular wave number of 0.1. This reduction starts at the fastest growing modes as they quickly scatter their resonant particles, and with time it extends to slower growing modes at higher parallel wave numbers.

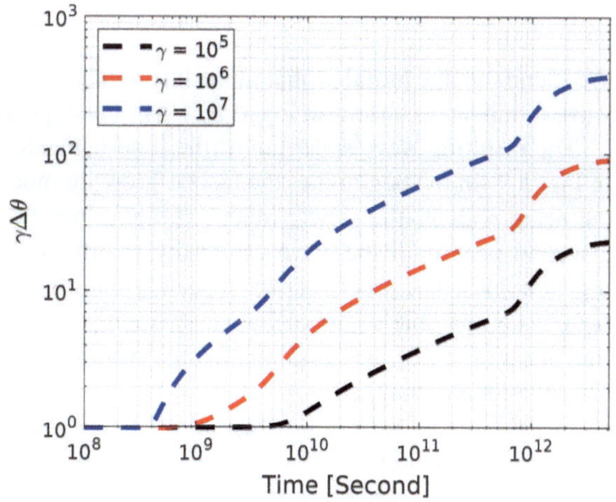

Fig. 5.7 The angular spread for different Lorentz factors of the beam as a function of time during the angular diffusion feedback simulation presented in Sect. 5.4.1

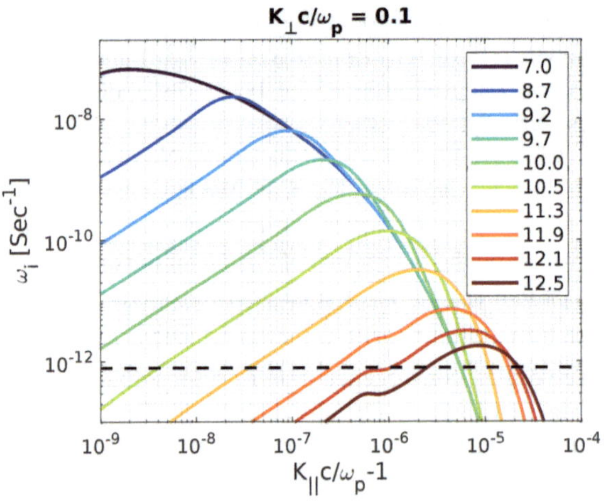

Fig. 5.8 Evolution of the linear growth rate of the instability for a fixed perpendicular wave number during the angular diffusion feedback simulation presented in Sect. 5.4.1. The black dashed line represents the collisional damping rate. Legend values are common logarithms of time in seconds. Throughout the simulation, we observed that the linear growth rate has maintained its initial profile with perpendicular wave numbers as in Fig. 5.5

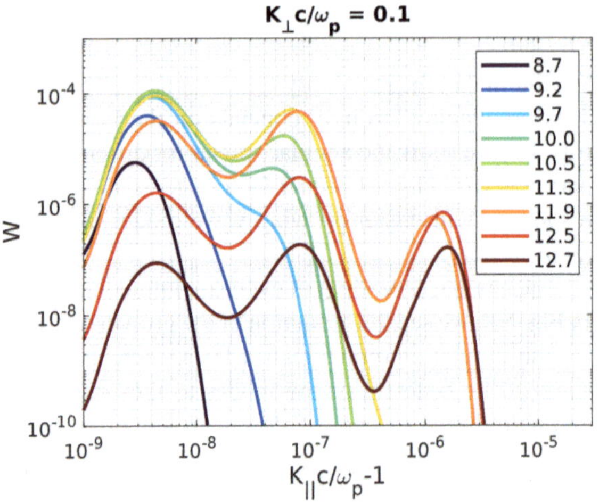

Fig. 5.9 The time evolution of the wave spectrum for fixed perpendicular wave number for the angular diffusion feedback simulation presented in Sect. 5.4.1. Here $W \equiv \frac{dW}{\frac{c}{\omega_p} d\frac{ck_\perp}{\omega_p} d\left(\frac{ck_\parallel}{\omega_p}-1\right)}$ in unites of eV cm^{-3}. Legend values are common logarithms of time in seconds

We found that the initial profile of the linear growth with respect to the perpendicular wave numbers, as shown in Fig. 5.5, doesn't change during the time evolution.

The resulting wave spectrum of the time-dependent linear growth rate is shown in Fig. 5.9. In the beginning, the fastest-growing modes form a spectral peak. Once the beam widens, the slower modes at higher parallel wave numbers start forming a second peak until the wave's intensity is sufficient to kick the beam particles to higher angles. The process keeps repeating until the linear growth rate becomes less than or comparable to the collisional damping rate (presented by the dashed black line in Fig. 5.8). By the time we stop the simulation at 5×10^{12} seconds, all modes are collisionally damped.

In Fig. 5.10, we show the diffusion coefficient, $D_{\theta\theta}$, at various times. The variation of the diffusion coefficient with beam angle, θ, closely resembles that of the wave

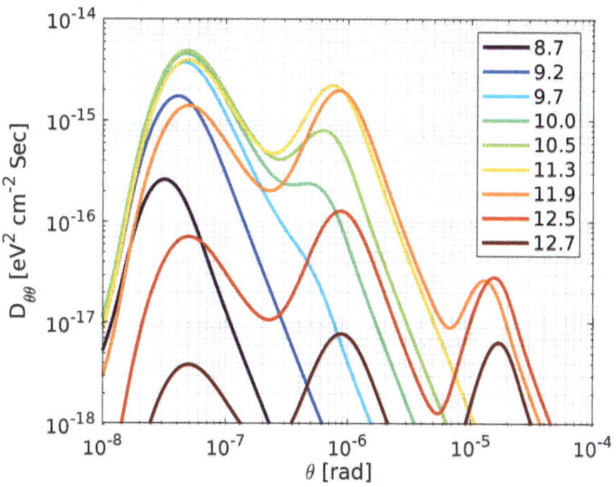

Fig. 5.10 The $D_{\theta\theta}$ for $\gamma = 10^6$ as a function of time in the angular diffusion feedback simulation presented in Sect. 5.4.1. Legend values are common logarithms of time in seconds

spectrum with parallel wave numbers, as represented in Fig. 5.9. This is due to the resonance relation between the beam angle and the parallel wave numbers $\theta = \left(\frac{ck_{\|}}{\omega_p} - 1\right) / \left(\frac{ck_{\perp}}{\omega_p}\right)$ in the regime $\left(\frac{ck_{\perp}}{\omega_p}\right)^2 \gg \left(\frac{ck_{\|}}{\omega_p} - 1\right)$. The emergence of the final prominent peak in the diffusion coefficient profile, occurring at angular values of approximately 10^{-5} radians, causes the observed increase in the angular spread rate after 7×10^{11} seconds, as seen in Fig. 5.7. We also see that around this time pairs with different Lorentz factors react to the same resonant unstable modes, because in the regime $\left(\frac{ck_{\perp}}{\omega_p}\right)^2 \gg \left(\frac{ck_{\|}}{\omega_p} - 1\right)$ waves with a certain parallel wave number are resonant with the beam particles at a certain angle, whatever their momentum.

We see in Fig. 5.7 that by the time the instability has saturated, the angular spread of pairs with Lorentz factor 10^6 has increased by around two orders of magnitudes, much more than the factor of ten reported by Perry and Lyubarsky [2]. The main reasons for this higher spread are the smaller collisional damping rate and the higher beam density we used.

We found that during the entire simulation time, the wave energy density never exceeded 10^{-3} of the beam energy density. This level of the wave intensity is lower than that needed for efficient operation of nonlinear Landau damping and the Modulation instability [4, 8, 15]. Therefore, the effect of these non-linear processes on the instability development might be minimal compared to that of the diffusive feedback on the beam.

We also calculated the total energy transferred from the beam to the waves by integrating the energy loss rate of the beam given in Eq. 5.38 over time. The result is given by the black dashed line in Fig. 5.11. We see that the beam lost less than 1% of its total initial energy by the time the instability development was saturated by the widening feedback. Those results suggest that the feedback widening severely limits the energy transfer from the beam to the waves. We explore whether this situation changes as we use different beam densities in Sect. 5.4.3.

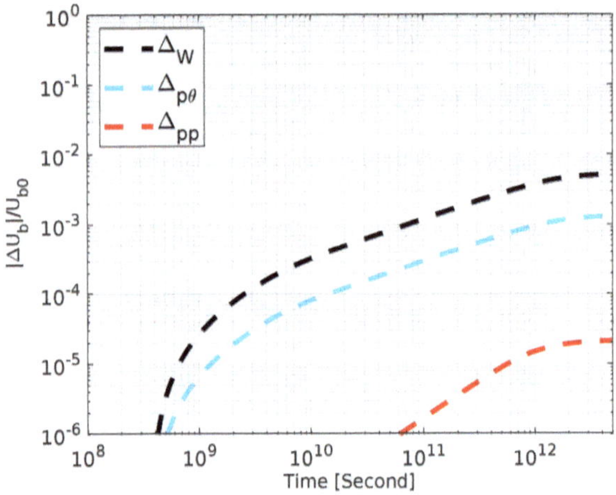

Fig. 5.11 The accumulated change in the beam energy during the angular diffusion feedback simulation presented in Sect. 5.4.1. The black dashed line (Δ_W) represents the beam energy fraction going into unstable wave growth. The dashed cyan line ($\Delta_{p\theta}$) and the dashed red line (Δ_{pp}) represent the fraction of the beam energy loss and gain due to the momentum diffusion by the $p\theta$ and the pp terms, respectively

Up to here, we only included the initially dominant term $D_{\theta\theta}$ of the right-hand side of Eq. 5.40. In the next section, we analyse the feedback of the other subdominant terms as the dominant $\theta\theta$ diffusion widens the pair beam.

5.4.2 2D Analysis of the Diffusion Equation

We analyse here the effect of the subdominant terms as the beam widens, using the time-dependent beam distribution that we numerically derived and discussed in the previous section.

For the momentum diffusion of the beam (third and fourth terms on the RHS of Eq. 5.40), we can calculate the energy loss or gain rate of the beam by inserting the corresponding time derivative of the beam distribution Eq. 5.40 in the total rate of change of the beam energy. After integrating by parts we get the following relation for $p\theta$ diffusion

$$\begin{aligned} \left.\frac{dU_b}{dt}\right|_{p\theta}(t) &= 2\pi m_e c^2 \int d\theta\theta \int dpp^2\gamma \left.\frac{df}{dt}\right|_{p\theta}(p,\theta,t) \\ &= -2\pi c \int d\theta\theta \int dpp D_{p\theta}\frac{\partial f}{\partial\theta}(p,\theta,t), \end{aligned} \tag{5.46}$$

and the following for pp diffusion

$$\left.\frac{dU_b}{dt}\right|_{pp}(t) = -2\pi c \int d\theta\theta \int dpp^2 D_{pp}\frac{\partial f}{\partial p}(p,\theta,t). \tag{5.47}$$

Looking at the overall sign of Eq. 5.46, we see that the diffusion $p\theta$ involves a global energy loss of the beam since $D_{p\theta}$ and the angular derivative are always

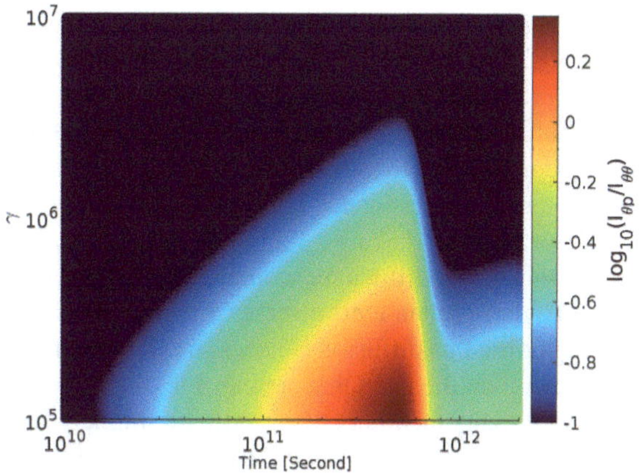

Fig. 5.12 The logarithm of the ratio of $I_{\theta p}$ (Eq. 5.49) and $I_{\theta\theta}$ (Eq. 5.48). The diffusion θp dominates over the term $(\theta\theta)$ in the orange and red areas with values higher than zero, while it contributes less than 10% in the dark blue area. The drop in the ratio just before the rim at 10^{12} seconds is due to the increase in the widening as a result of wave growth outside the initial resonance region. After 10^{12} seconds, the collisional damping effectively damps the waves, and the impact of both terms declines

negative. We also found that the dominant feedback of the diffusion pp is an energy gain of the beam, as D_{pp} is always positive and the beam distribution function declines for $\gamma \gtrsim 10^5$ (see Fig. D.1).

Integrating Eqs. 5.46 and 5.47 over the simulation time and dividing by the total beam initial energy, we see in Fig. 5.11 the accumulated fraction of the beam energy lost and gained. We observe that $p\theta$ diffusion could eliminate only around 0.1% of the beam total energy by the end of the simulation whereas pp diffusion increases it by a negligible fraction. Therefore, it is evident that the cumulative effect of the diffusive momentum flux on the beam is insignificant compared to the scattering.

Now, we proceed to the analysis of the second term on the RHS of Eq. 5.40, θp diffusion. This diffusion involves angular flux as the $\theta\theta$ diffusion, but it can result in both the narrowing and widening of the beam depending on the beam momentum gradient. Pairs with negative momentum gradient, $\gamma > 10^5$, experience narrowing whereas the ones with positive momentum gradient, $\gamma < 10^5$, experience a widening. In Fig. 5.12, we have compared the normalized angular integral of the absolute rate of change of the $\theta\theta$ diffusion for a certain beam Lorentz factor

$$I_{\theta\theta} = \int d\cos\theta \left|\frac{df}{dt}\bigg|_{\theta\theta}\right| = \int d\cos\theta \left|\frac{1}{p^2\theta}\frac{\partial}{\partial\theta}\left(\theta D_{\theta\theta}\frac{\partial f}{\partial\theta}\right)\right|, \tag{5.48}$$

with that one of the θp diffusion

$$I_{\theta p} = \int d\cos\theta \left|\frac{df}{dt}\bigg|_{\theta p}\right| = \int d\cos\theta \left|\frac{1}{p\theta}\frac{\partial}{\partial\theta}\left(\theta D_{\theta p}\frac{\partial f}{\partial p}\right)\right|. \tag{5.49}$$

It is noticeable in Fig. 5.12 that the ratio of $I_{\theta p}$ and $I_{\theta\theta}$ increases gradually until it drops after 7×10^{11} s. The reason for the increase is that the diffusive flux of $\theta\theta$ decreases as the beam profile flattens, while the diffusive flux of θp remains relatively constant as the momentum gradients are not impacted by the beam broadening. The drop after 7×10^{11} seconds is due to the increase of diffusion $\theta\theta$ by the accumulated wave density outside the initial resonance region that we discussed in the previous section.

In Fig. 5.12 we see that the θp diffusion becomes dominant for Lorentz factors less than 10^6 at times much earlier than their inverse Compton cooling time ($\gtrsim 10^{13}$ s). For these particles including this diffusion is necessary. However, there is a minimal impact of the θp diffusion on the pairs that are capable of giving IC emission in the detectable GeV band (Lorentz factors of 10^6 or slightly higher). This indicates that θp diffusion might not impact the GeV-scale cascade emission as strongly as the $\theta\theta$ diffusion does.

5.4.3 Parameters Dependence

In the simulation discussed in Sect. 5.4.1, we used a fiducial pair beam density at a distance of 50 Mpc from the blazar, $3 \times 10^{-22}\text{cm}^{-3}$ [3]. However, the beam density changes under different conditions, such as varying the distance from the source, changing the source's luminosity, or using different EBL models in the calculations. Here we vary the beam density using the same setup as in Sect. 5.4.1 and investigate its impact on our results.

In Fig. 5.13, we see the fraction of the beam energy lost by the instability for different beam densities. As the beam density is increased, the instability develops earlier and takes more energy from the beam. However, the beam lost only 2% even

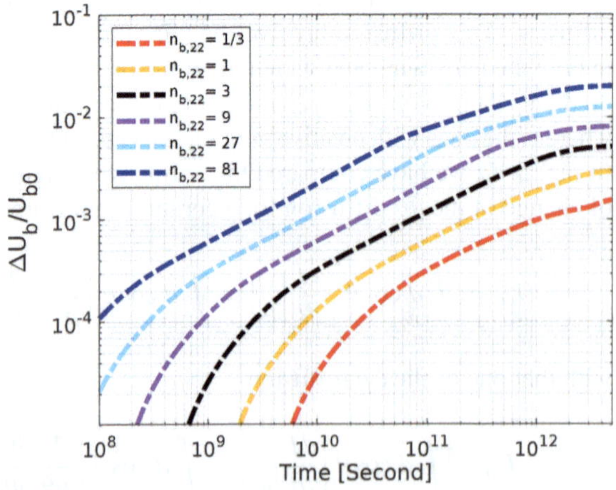

Fig. 5.13 The accumulated fraction of the beam energy lost as a function of time due to the wave growth during the angular diffusion simulations feedback with different values of the beam density. All the values are in units of 10^{-22} cm^{-3}

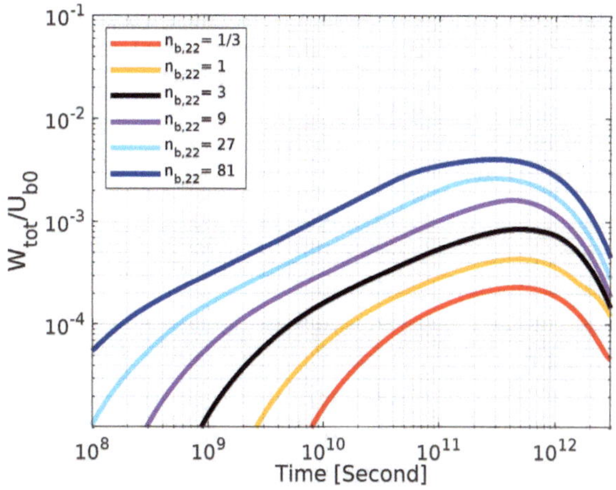

Fig. 5.14 The total energy density contained in the electric fields with time during the angular diffusion simulations feedback with different values of the beam density. The total energy density starts to decrease around 5×10^{11} seconds due to the decay of the waves by the collisional damping

for a very high beam density, 8×10^{-21} cm^{-3}. Therefore, the fundamental physical behaviour of the system remains consistent, the beam experiences expansion with a negligible energy loss of its initial energy as the instability is saturated by the beam expansion.

In Fig. 5.14, we plot the time evolution of the total electric field energy density, Eq. 5.37, for different beam densities. We noticed that the wave intensity during the simulations with beam density higher than 10^{-21}cm^{-3} has exceeded the threshold for the non-linear modulation instability that is around 10^{-3}. We didn't include any of the wave spectrum non-linear processes in our calculations. Those processes will impose further restrictions on the growth of the unstable modes.

We observed that the angular spread increased by a factor of 1.5 when the beam density was inflated by a factor of three (see table 5.1). This scaling can be attributed to the fact that the linear growth rate is linearly proportional to the pair beam density and inversely proportional to the square of the beam angular spread, $\omega_i \propto \frac{n_b}{\Delta\theta^2}$. Therefore, increasing the beam density by a factor C requires an increase in the angular spread by a factor of approximately $\sqrt{C}$ to maintain the reduction of the linear growth rate to the collisional damping rate at the time when the instability has saturated.

In the remainder of this section, we will discuss the influence of the cut-off energy in the intrinsic gamma-ray spectrum on the results of Sect. 5.4.1. Vafin et al. [3] used an intrinsic power-law gamma-ray spectrum with a step function cut-off at the energy of 50 TeV. However, in the end, they used the attenuated gamma-ray spectrum at a distance of 50 Mpc to calculate the accumulated pair beam spectrum over a certain path length. At a distance of 50 Mpc from the blazar, the majority of gamma rays with energies higher than 10 TeV have already been absorbed. The mean free path for a gamma-ray with energy E_γ for pair production with the EBL photons is given by

$$\lambda_{\gamma\gamma} \approx 80\,(1+z)^{-\xi} \left(\frac{E_\gamma}{10\text{TeV}}\right)^{-1} \text{Mpc}, \tag{5.50}$$

Table 5.1 The angular spread, $\Delta\theta_F$, for different beam Lorentz factors, γ, at the end time, 5×10^{12} seconds, of the $\theta\theta$ diffusion simulation presented in Sect. 5.4.1 with different beam densities, n_b. We see here clearly that the angular spread increased by a factor of 1.5 when the beam density was inflated by a factor of three

$\gamma = 10^5$				
$n_b[10^{-22}\text{cm}^{-3}]$	$\Delta\theta_F$ [rad]	Ratio		
3	2.26×10^{-4}	1.53		
9	3.46×10^{-4}		1.51	
27	5.21×10^{-4}			1.52
81	7.91×10^{-4}			
$\gamma = 10^6$				
$n_b[10^{-22}\text{cm}^{-3}]$	$\Delta\theta_F$ [rad]		Ratio	
3	8.96×10^{-5}	1.52		
9	1.36×10^{-4}		1.49	
27	2.03×10^{-4}			1.5
81	3.00×10^{-4}			
$\gamma = 10^7$				
$n_b[10^{-22}\text{cm}^{-3}]$	$\Delta\theta_F$ [rad]	Ratio		
3	3.63×10^{-5}	1.51		
9	5.48×10^{-5}		1.48	
27	8.10×10^{-5}			1.47
81	1.19×10^{-4}			

where $\xi = 4.5$ and $\xi = 0$ for redshifts of $z \leq 1$ and $z > 1$, respectively [23, 24]. Therefore, any cut-off energy above the 10-TeV threshold will have only a minimal impact.

5.4.4 *Simulation with Injection*

In Sect. 5.4.1, we found that the instability growth is severely reduced by beam broadening to the point that it cannot be isolated from the production and the cooling rates of the beam. In this section, we include the pair creation rate in the evolution equation of the beam.

The beam distribution found in Vafin et al. [3] was calculated as the accumulation of pairs over the path length of 7.7×10^{12} light-seconds, using a constant production rate Q_{ee}. We added this production rate to the beam evolution equation along with the dominant $\theta\theta$ diffusion term,

$$\frac{\partial f(p,\theta)}{\partial t} = \frac{1}{p^2\theta}\frac{\partial}{\partial\theta}\left(\theta D_{\theta\theta}\frac{\partial f}{\partial\theta}\right) + Q_{ee}. \tag{5.51}$$

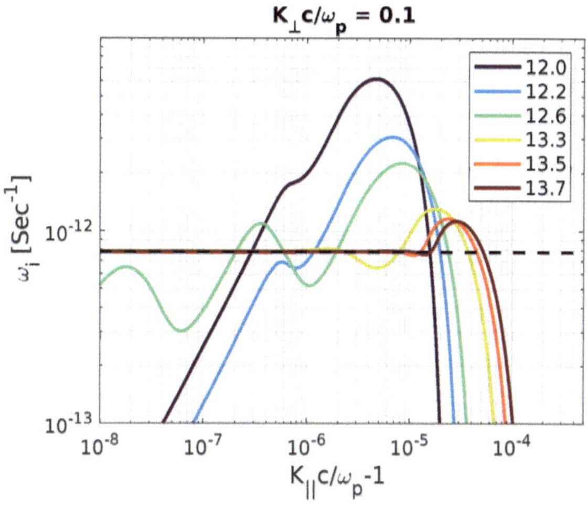

Fig. 5.15 Evolution of the linear growth rate of the instability for a fixed perpendicular wave number during the simulation with continuous pair injection presented in Sect. 5.4.4. The black dashed line represents the collisional damping rate. Legend values are common logarithms of time in seconds. We see the linear growth rate eventually balance the collisional damping rate across the wave spectrum after the time 10^{13} seconds. We see also that the resonance region keeps expanding due to the ongoing expansion of the beam particles to higher opening angles

Using the same simulation setup as described in Sect. 5.4.1, we numerically solved the coupled system of the evolution equations (Eqs. 5.51 and 5.35). For times much less than 10^{13} seconds, we found essentially the same behaviour of the system as without injection. After 10^{13} s, a new quasi-steady state of the beam distribution and the waves spectrum emerges.

The creation of highly focused pairs with beam angles of the order γ^{-1} increases the linear growth rate at wave numbers in resonance with these particles. Ultimately, this leads to a quasi-equilibrium of the wave spectrum and the beam distribution. On the wave side, the linear growth rate and the collisional damping rate balance across the resonant wave numbers (see Fig. 5.15), resulting in a steady-state wave spectrum as shown in Fig. 5.16 and a steady-state diffusion coefficients as well as shown in Fig. 5.17. On the beam side, the diffusive scattering compensates the pair production, keeping the beam expanding as shown in the angular profile of pairs with a Lorentz factor of 10^6 in Fig. 5.18. This ongoing expansion of the beam extends the unstable modes to higher parallel wave numbers as shown in Fig. 5.16.

We have stopped the simulation after 5×10^{13} s, which corresponds to the IC cooling time of pairs with a Lorentz factor of 10^6. By this time the pairs have experienced a diffusive deflection up to angles of around 4×10^{-4} radians. This deflection results in an arrival time delay of the secondary GeV-band photons emitted by those pairs [25]. The arrival time delay of secondary gamma rays emitted by pairs that have undergone a deflection by an angle of $\Delta\theta$ from the primary gamma-ray propagation direction is given by the following formula

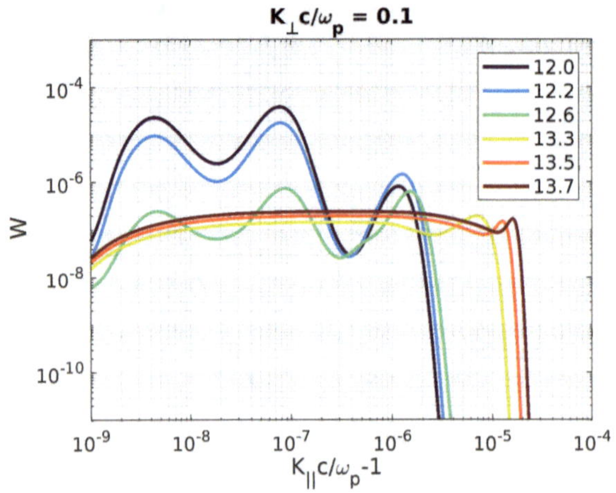

Fig. 5.16 The wave spectrum as a function of time during the injection simulation presented in Sect. 5.4.4. Legend values are common logarithms of time in seconds. We see the steady-state diffusion coefficients emerging after the time 10^{13} s

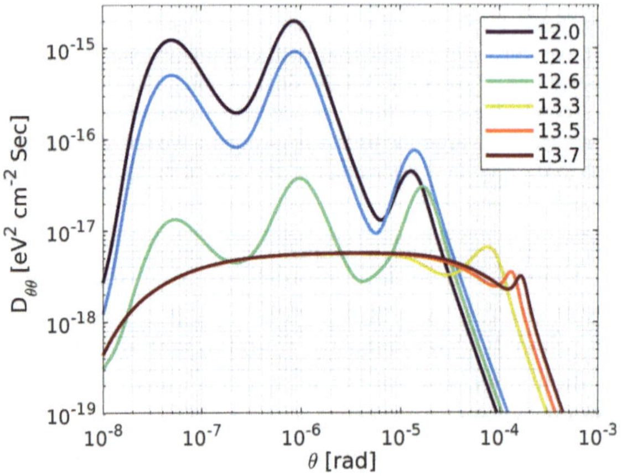

Fig. 5.17 The time evolution of $D_{\theta\theta}$ for $\gamma = 10^6$ during the simulation with continuous pair injection presented in Sect. 5.4.4. Legend values are common logarithms of time in seconds

$$\Delta t_{\text{delay}} \simeq \frac{\Delta\theta^2}{2} \frac{D_c D_b - D_c^2}{c D_b}, \tag{5.52}$$

where D_c is the distance between the emitting pairs and the blazar and D_b is the distance between the blazar and the Earth. Given that for our simulation setup $D_c = 50\,\text{Mpc}$ and $D_b = 720\,\text{Mpc}$ for the fiducial $z = 0.15$, the formula reduces to $\Delta t_{\text{delay}} = \Delta\theta^2 \times 7.6 \times 10^7$ years. Hence, the deflection of pairs with Lorentz factor of 10^6 by 4×10^{-4} radians implies a time delay of around 10 years for the GeV-scale cascade emission produced at the distance 50 Mpc from the source. Calculating the deflection at different distances from the source is needed to find the impact on the observed cascade emission. This is beyond the scope of this paper and will be covered in future works.

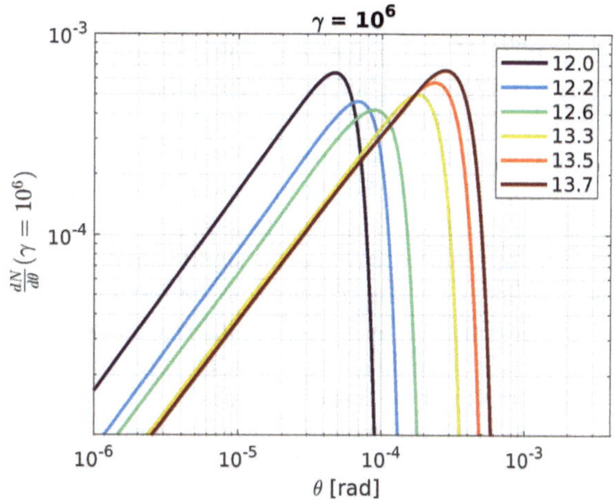

Fig. 5.18 The angular profile of pairs with Lorentz factor of 10^6 during the injection simulation presented in Sect. 5.4.4. Legend values are common logarithms of time in seconds. We see that around the IC cooling time of 5×10^{13}s for those pairs, they have been deflected by around 4×10^{-4} radians which yields a time delay of around 10 years for the GeV cascade

As of the previous simulation in Sect. 5.4.1, we also found here that the wave energy density never exceeded 10^{-3} of the total beam energy density, keeping the wave intensity at levels lower than what is needed for the non-linear processes to operate efficiently, again justifying their neglect.

It is crucial to emphasize that the calculations presented in this paper have not incorporated IC cooling. The inclusion of the IC cooling is important for a comprehensive understanding of the associated physical implications on the arrival time distribution of GeV-scale cascade emissions, and it is part of our future research plans.

5.5 Summary

We solved the Fokker-Planck diffusion equation for the beam distribution function, coupled with the linear evolution equation of the plasma-wave spectrum. As the initial condition for the beam, we used the realistic two-dimensional beam distribution computed by Vafin et al. [3] for a distance of 50 Mpc from the blazar. Initially, the dominant feedback is angular broadening of the beam, stemming from the scattering of the beam particles by the excited waves. As the instability widens the beam, the instability growth rate is severely reduced, leading at the end to a negligible energy transfer from the beam to the plasma waves. These findings align with a recent study on instability feedback [2].

Using the 2D time-dependent beam profile evolving by the predominant angular diffusion, we found that momentum diffusion does not have any significant impact on the beam. However, we found that another angular diffusion term, which is initially negligible, might become relevant and may narrow the beam particles with Lorentz factors between 10^5 and 10^6. Therefore, including this term in the feedback calculations is necessary for a comprehensive understanding of the instability impact on those pairs. However, the GeV-scale cascade is emitted by pairs with Lorentz factors of 10^6 or slightly higher, and so the impact of this term on the GeV-scale secondary cascade might be limited.

In our analysis, we neglected non-linear wave interactions in the evolution of the wave spectrum. For beam density lower than 10^{-21} cm^{-3}, we found that the cumulative energy density of the electric field fluctuations remains below the critical thresholds required to trigger the significant impacts of the non-linear processes, such as non-linear Landau damping or the modulation instability. However, for higher beam densities the wave energy density exceeded the threshold for the non-linear modulation instability. Those non-linear processes would impose further restrictions on the growth of the unstable modes.

Lastly, we have included the continuous TeV pairs production in the Fokker-Planck diffusion equation. Unlike the previous simulation discussed in Sect. 5.4.1, in this particular configuration, the unstable modes do not decay after the beam has expanded but saturate at a finite amplitude. The wave spectrum reaches a quasi-equilibrium across the wave numbers resonant with the beam injection angles. The beam particles experience persistent scattering under the diffusive feedback of this steady-state wave spectrum. Then, beam particles with Lorentz factors of 10^6 scatter up to angles of around 4×10^{-4} radians within their IC cooling time. This results in a time delay of around 10 years in the arrival of the secondary GeV-scale cascade, assuming pairs at a distance of 50 Mpc from a blazar that is 720 Mpc away from Earth. We expect that this estimate depends on the beam density that varies along the propagation distance and with source luminosity.

In the end, calculating the broadening at more points along the beam propagation is needed to understand the impact of the instability broadening on the GeV-scale cascade emission. Also, it's essential to include the inverse Compton cooling in the beam distribution evolution equation to understand the long-term time evolution of the beam-wave system.

References

1. Mahmoud Alawashra and Martin Pohl. Nonlinear feedback of the electrostatic instability on the blazar-induced pair beam and gev cascade. *The Astrophysical Journal*, 964(1):82, mar 2024. https://doi.org/10.3847/1538-4357/ad24ea
2. Roy Perry and Yuri Lyubarsky. The role of resonant plasma instabilities in the evolution of blazar-induced pair beams. *MNRAS*, 503(2):2215–2228, May 2021. https://doi.org/10.1093/mnras/stab324
3. S. Vafin, I. Rafighi, M. Pohl, and J. Niemiec. The electrostatic instability for realistic pair distributions in blazar/ebl cascades. *The Astrophysical Journal*, 857(1):43, Apr 2018. ISSN 1538-4357. https://doi.org/10.3847/1538-4357/aab552
4. S. Vafin, P. J. Deka, M. Pohl, and A. Bohdan. Revisit of nonlinear landau damping for electrostatic instability driven by blazar-induced pair beams. *The Astrophysical Journal*, 873(1):10, feb 2019. https://doi.org/10.3847/1538-4357/ab017b

5. Crank J, Nicolson P (1947) A practical method for numerical evaluation of solutions of partial differential equations of the heat-conduction type. Mathematical Proceedings of the Cambridge Philosophical Society 43(1):50–67. https://doi.org/10.1017/S0305004100023197
6. William H Press. *Numerical recipes 3rd edition: The art of scientific computing*. Cambridge university press, 2007
7. Avery E. Broderick, Philip Chang, and Christoph Pfrommer. THE COSMOLOGICAL IMPACT OF LUMINOUS TeV BLAZARS. i. IMPLICATIONS OF PLASMA INSTABILITIES FOR THE INTERGALACTIC MAGNETIC FIELD AND EXTRAGALACTIC GAMMA-RAY BACKGROUND. *The Astrophysical Journal*, 752(1):22, may 2012. https://doi.org/10.1088/0004-637x/752/1/22
8. Francesco Miniati and Andrii Elyiv. RELAXATION OF BLAZAR-INDUCED PAIR BEAMS IN COSMIC VOIDS. *The Astrophysical Journal*, 770(1):54, may 2013. https://doi.org/10.1088/0004-637x/770/1/54
9. A. Bret, L. Gremillet, and M. E. Dieckmann. Multidimensional electron beam-plasma instabilities in the relativistic regime. *Physics of Plasmas*, 17(12):120501, 2010b. https://doi.org/10.1063/1.3514586
10. A. Bret, M. C. Firpo, and C. Deutsch. Electromagnetic instabilities for relativistic beam-plasma interaction in whole k space: Nonrelativistic beam and plasma temperature effects. *Phys. Rev.*, 72(1):016403, July 2005. https://doi.org/10.1103/PhysRevE.72.016403
11. I. Rafighi, S. Vafin, M. Pohl, and J. Niemiec. Plasma effects on relativistic pair beams from TeV blazars: PIC simulations and analytical predictions. *Astronomy &Astrophysics*, 607:A112, November 2017. https://doi.org/10.1051/0004-6361/201731127
12. Philip Chang, Avery E. Broderick, Christoph Pfrommer, Ewald Puchwein, Astrid Lamberts, Mohamad Shalaby, and Geoffrey Vasil. The Linear Instability of Dilute Ultrarelativistic e $^{\pm}$ Pair Beams. *ApJ*, 833(1):118, December 2016. https://doi.org/10.3847/1538-4357/833/1/118
13. B.N. Breizman. Reviews of plasma physics. *Reviews of Plasma Physics*, 15, January 1990
14. R. Schlickeiser, S. Krakau, and M. Supsar. PLASMA EFFECTS ON FAST PAIR BEAMS. II. REACTIVE VERSUS KINETIC INSTABILITY OF PARALLEL ELECTROSTATIC WAVES. *The Astrophysical Journal*, 777(1):49, oct 2013. https://doi.org/10.1088/0004-637x/777/1/49
15. Philip Chang, Avery E. Broderick, Christoph Pfrommer, Ewald Puchwein, Astrid Lamberts, and Mohamad Shalaby. THE EFFECT OF NONLINEAR LANDAU DAMPING ON ULTRA-RELATIVISTIC BEAM PLASMA INSTABILITIES. *The Astrophysical Journal*, 797(2):110, dec 2014. https://doi.org/10.1088/0004-637x/797/2/110
16. S F Tigik, P H Yoon, and L F Ziebell. Bremsstrahlung emission and collisional damping rate for langmuir waves. *Plasma Physics and Controlled Fusion*, 61(12):125008, oct 2019. https://doi.org/10.1088/1361-6587/ab4aad
17. P. H. Yoon, L. F. Ziebell, E. P. Kontar, and R. Schlickeiser. Weak turbulence theory for collisional plasmas. *Phys. Rev. E*, 93:033203, Mar 2016. https://doi.org/10.1103/PhysRevE.93.033203
18. Shalaby Mohamad, Broderick Avery E, Chang Philip, Pfrommer Christoph, Puchwein Ewald, Lamberts Astrid (2020) The growth of the longitudinal beam–plasma instability in the presence of an inhomogeneous background. Journal of Plasma Physics 86(2):535860201. https://doi.org/10.1017/S0022377820000215
19. R. Schlickeiser, D. Ibscher, and M. Supsar. PLASMA EFFECTS ON FAST PAIR BEAMS IN COSMIC VOIDS. *The Astrophysical Journal*, 758(2):102, oct 2012. https://doi.org/10.1088/0004-637x/758/2/102
20. Lorenzo Sironi and Dimitrios Giannios. RELATIVISTIC PAIR BEAMS FROM TeV BLAZARS: A SOURCE OF REPROCESSED GeV EMISSION RATHER THAN INTERGALACTIC HEATING. *The Astrophysical Journal*, 787(1):49, may 2014. https://doi.org/10.1088/0004-637x/787/1/49
21. B.N. Brejzman and D.D. Ryutov. Powerful relativistic electron beams in a plasma and in a vacuum (theory). *Nuclear Fusion*, 14(6):873–907, dec 1974. https://doi.org/10.1088/0029-5515/14/6/012

22. L. I. Rudakov. Collective Slowing Down of an Intense Beam of Relativistic Electrons in a Dense Plasma Target. *Soviet Journal of Experimental and Theoretical Physics*, 32:1134, January 1971
23. Kneiske, T. M., Bretz, T., Mannheim, K., and Hartmann, D. H. Implications of cosmological gamma-ray absorption - ii. modification of gamma-ray spectra. *A&A*, 413(3):807–815, 2004. https://doi.org/10.1051/0004-6361:20031542
24. A. Neronov and D. V. Semikoz. Sensitivity of γ-ray telescopes for detection of magnetic fields in the intergalactic medium. *Phys. Rev. D*, 80:123012, Dec 2009. https://doi.org/10.1103/PhysRevD.80.123012
25. A. Neronov and D. V. Semikoz. Sensitivity of γ-ray telescopes for detection of magnetic fields in the intergalactic medium. *Physical Review D*, 80(12):123012, December 2009. https://doi.org/10.1103/PhysRevD.80.123012

Chapter 6
Conclusions and Outlook

Very high-energy gamma rays emitted by blazars produce relativistic pair beams in the intergalactic medium (IGM) after interacting with the extragalactic background light (EBL). Those pairs are expected to produce a detectable inverse Compton cascade in the GeV energy band that is missing in the observations of the gamma-ray telescope Fermi-LAT. The suppression of this secondary cascade implies either the deflection of the pair beam by intergalactic magnetic fields (IGMFs) [1] or, alternatively, an energy loss of the beam due to the beam-plasma instability [2]. The possible role of the electrostatic instability of blazar-induced pair beams has been studied both analytically and numerically but no definite answer has been reached in the literature.

It's evident that the linear growth rate of the instability for narrow realistic blazar-induced beams, $\Delta\theta \sim \gamma^{-1}$, is much faster than the inverse Compton (IC) cooling rate on the CMB photons [3]. What is not clear is whether the non-linear evolution of the instability succeeds at draining a significant fraction of the beam energy by the IC cooling time [4–6]. We have reviewed and investigated this problem in the Sects. 3.5 and 3.6. We found that the modulation instability is a faster saturation process than the nonlinear Landau damping. Combining these two non-linear processes in a simplified model we found in Sect. 3.6 that the energy density of the resonant plasma waves saturates at the threshold of the Modulation instability draining the beam energy faster than the IC cooling.

Assuming that the non-linear saturation of the instability succeeds at draining the beam energy before the IC scattering on the CMB, there are still two significant effects that have not been included in this picture. The first one is the magnetization of the intergalactic medium and the second one is the instability feedback on the beam. In this thesis, we have studied those two effects in detail. In Chap. 4, and separately published as [7], we have investigated the effects of tangled weak IGMFs with small correlation lengths on the electrostatic instability. Whereas in Chap. 5, and separately published as [8], we have simulated the beam-plasma system considering the feedback of the instability on the pair beam in the quasi-linear regime.

M. S. A. Alawashra, *Plasma Instabilities of TeV Pair Beams Induced by Blazars*,
Springer Theses, https://doi.org/10.1007/978-3-031-98568-3_6

In Chap. 4, we have studied the influence of weak IGMFs on the efficient energy loss of the beam by the instability. We found that tangled weak IGMFs, with correlation lengths smaller than the pairs' energy loss lengths, deflect the pair beam particles stochastically increasing the beam angular spread. The increment of the angular spread decreases the instability growth rate, which in turn reduces the associated energy loss rate. Considering the instability energy loss time found in Vafin et al. [3], We found the values of the intergalactic magnetic fields strength B_{IGM} and correlation length λ_B sufficient to suppress the instability energy loss rate to the level of that one of the IC cooling.

Those values of B_{IGM}-λ_B that suppress the instability lay at IGMF strengths that are three orders of magnitude lower than the estimated IGMF strengths needed to suppress the secondary cascade by ten years of time delay due to the magnetic deflection [9]. In the region of the B_{IGM}-λ_B parameter space that suppresses the instability and that is below the time delay lower limit, neither the beam-plasma instability nor the intergalactic magnetic field deflection can explain the absence of cascade emission in the spectra of some TeV blazars, and so this parameter space region can be excluded unless there is a third mechanism that suppresses the GeV-band cascade.

In conclusion, we can exclude an IGMF strength within the three orders of magnitude below the limit above which magnetic deflection imposes a significant time delay of the cascade (ten years). This leaves a conditionally new allowed region for the IGMF that lies below 10^{-16} Gauss at $\lambda_{\text{B}} \approx 10^{-8}$ Mpc, and the constraint on the IGMF strength is tighter than that for both larger correlation lengths (due to the instability suppression) and smaller correlation lengths (due to the MHD turbulent decay). The condition for this allowed region is the instability-efficient energy draining of the beam. Perry and Lyubarsky [10] argued otherwise of this condition and that the instability feedback broadens the beam predominantly with negligible energy loss.

In Chap. 5, we have explored the feedback of the instability on a realistic two-dimensional beam distribution, unlike the simplified one-dimensional distribution used in Perry and Lyubarsky [10]. The feedback of the instability is described by a Fokker-Planck diffusion in both beam momentum and angles. Beam narrowing and momentum diffusion were neglected in the one-dimensional treatment of Perry and Lyubarsky [10] considering only the initially dominant beam broadening. Our goal was to check the validity of this approximation using the information from the two-dimensional beam profile evolving under the influence of the initially dominant broadening feedback.

We solved the Fokker-Planck beam diffusion equation including the instability dominant broadening feedback, coupled with the linear evolution of the unstable wave modes. We found that the angular broadening of the beam, stemming from the scattering of the beam particles by the unstable waves, severely reduced the instability growth rate leading at the end to a negligible energy transfer from the beam to the unstable waves. We observed by the time the instability has saturated, an increase in the beam angular spread by a factor of a hundred for pairs with Lorentz factor of 10^6, this is much more than the factor of ten reported by Perry and Lyubarsky [10].

The main reasons for this higher spread are the more accurate and smaller collisional damping rate and the higher beam density we used.

We found using the evolving 2D time-dependent beam profile under the predominant angular diffusion feedback that the instability momentum diffusion is incapable of any significant impact on the beam. However, we observed that a different angular diffusion term, initially negligible, may become relevant and lead to the narrowing of the beam particles with Lorentz factors ranging from 10^5 to 10^6. Consequently, incorporating this term into the feedback calculations becomes essential for a comprehensive comprehension of the instability's physical impact on these pairs. However, the emission of a GeV-scale cascade is expected to be emitted predominantly by pairs with Lorentz factors of 10^6 or slightly higher, and thus, the effect of this term on the observable GeV-scale cascade is anticipated to be limited.

Finally, we have upgraded the beam-plasma quasi-linear simulation by including a source term of the continuous pairs production in the IGM due to the gamma rays annihilation with the EBL photons along with the dominant Fokker-Planck diffusion feedback of the instability. We found that in this case, the unstable modes do not decay completely after the beam expansion but saturate at a finite amplitude due to the continuous injection of the pairs with focused angles of γ^{-1}. In this new quasi-steady state, the wave spectrum reaches a steady state equilibrium across the wave numbers that are resonant with the beam injection angles. The beam particles experience persistent scattering under the diffusive feedback of this steady-state wave spectrum.

We found that beam particles with Lorentz factors of 10^6 scatter up to angles of around 4×10^{-4} radians by their IC cooling time. This deflection of the pairs by the instability scattering feedback would translate into an arrival time delay of the secondary cascade emission to Earth. We found that the time delay of a GeV-scale secondary cascade emitted by scattered pairs at a distance of 50 Mpc from a blazar that is 720 Mpc away from Earth is of the order of 10 years. Many improvements can be made to our calculations for a solid understanding of the observable effects of this new intrinsic scattering of the cascade by the instability feedback.

In our analysis, we simulated the beam-plasma system starting with the accumulated beam distribution at a distance of 50 Mpc from the blazar [3]. A proper understanding of the observable effects of the instability feedback requires calculating the broadening of the beam at different points along the distance between the blazar and Earth. The resulting angular spread would scale in this case with the beam density that varies along the propagation distance. The beam broadening estimate could also change due to the evolving momentum profile of the beam along the propagation distance.

Another important improvement would be to include the inverse Compton cooling in the beam transport equation. Including this term is necessary to understand the long-term time evolution of the new beam-spectrum quasi-steady state. One also can include the other initially subdominant angular diffusion θp term that dominates for pairs with Lorentz factors less than 10^6 after the expansion of the beam. This other instability feedback could scatter those beam particles inward affecting the beam broadening.

In the instability feedback calculations, we have neglected non-linear wave interactions in the evolution of the wave spectrum. This is valid as long as the energy density of the electric field fluctuations remains below the critical thresholds required to trigger the significant impacts of the non-linear processes, such as non-linear Landau damping and modulation instability. We found that this condition breaks when the beam densities are higher than 10^{-21} cm^{-3}. Incorporating the non-linear wave interaction terms in the wave evolution equation would be another improvement of our model. Those non-linear processes would impose further restrictions on the growth of the unstable modes.

Ultimately, our findings in this thesis suggest that the efficient energy loss of the blazar-induced pair beam by the instability could be limited by the beam broadening due to weak IGMFs and due to the instability feedback. This hints that the lack of GeV bumps in the blazar spectra might not be due to the beam energy loss by the plasma instability. However, we found that the instability can modify the beam to a new steady state with higher angular spreads affecting the arrival time of the GeV cascade. Several advancements are yet to be made to have a comprehensive understanding of the overall impact of this on the GeV cascade flux, such as adding the IC cooling of the pairs, simulating the beam-plasma system at different points in the IGM and including the non-linear processes in the wave evolution equation.

References

1. Andrii Neronov and Ievgen Vovk. Evidence for Strong Extragalactic Magnetic Fields from Fermi Observations of TeV Blazars. *Science*, 328(5974):73, April 2010. https://doi.org/10.1126/science.1184192
2. Avery E. Broderick, Philip Chang, and Christoph Pfrommer. THE COSMOLOGICAL IMPACT OF LUMINOUS TeV BLAZARS. i. IMPLICATIONS OF PLASMA INSTABILITIES FOR THE INTERGALACTIC MAGNETIC FIELD AND EXTRAGALACTIC GAMMA-RAY BACKGROUND. *The Astrophysical Journal*, 752(1):22, may 2012. https://doi.org/10.1088/0004-637x/752/1/22
3. S. Vafin, I. Rafighi, M. Pohl, and J. Niemiec. The electrostatic instability for realistic pair distributions in blazar/ebl cascades. *The Astrophysical Journal*, 857(1):43, Apr 2018. ISSN 1538-4357. https://doi.org/10.3847/1538-4357/aab552
4. Miniati Francesco, Elyiv Andrii (May 2013) RELAXATION OF BLAZAR-INDUCED PAIR BEAMS IN COSMIC VOIDS. The Astrophysical Journal 770(1):54
5. Chang Philip, Broderick Avery E, Pfrommer Christoph, Puchwein Ewald, Lamberts Astrid, Shalaby Mohamad (Dec 2014) THE EFFECT OF NONLINEAR LANDAU DAMPING ON ULTRARELATIVISTIC BEAM PLASMA INSTABILITIES. The Astrophysical Journal 797(2):110
6. Vafin S, Deka PJ, Pohl M, Bohdan A (Feb 2019) Revisit of nonlinear landau damping for electrostatic instability driven by blazar-induced pair beams. The Astrophysical Journal 873(1):10
7. Alawashra Mahmoud, Pohl Martin (Apr 2022) Suppression of the TeV Pair-beam-Plasma Instability by a Tangled Weak Intergalactic Magnetic Field. The Astrophysical Journal 929(1):67

8. Alawashra Mahmoud, Pohl Martin (Mar 2024) Nonlinear feedback of the electrostatic instability on the blazar-induced pair beam and gev cascade. The Astrophysical Journal 964(1):82
9. M. Ackermann, M. Ajello, L. Baldini, J. Ballet, G. Barbiellini, D. Bastieri, R. Bellazzini, E. Bissaldi, R. D. Blandford, E. D. Bloom, and et al. The search for spatial extension in high-latitude sources detected by the fermi large area telescope. *The Astrophysical Journal Supplement Series*, 237(2):32, Aug 2018. ISSN 1538-4365. https://doi.org/10.3847/1538-4365/aacdf7
10. Roy Perry and Yuri Lyubarsky. The role of resonant plasma instabilities in the evolution of blazar-induced pair beams. *MNRAS*, 503(2):2215–2228, May 2021. https://doi.org/10.1093/mnras/stab324

Appendix A
Beam Angular Distribution

A.1 Beam Distribution Function with IGMF

Consider a magnetic field with a constant magnitude that arbitrarily and abruptly changes its direction every correlation length, λ_B, along the beam propagation line. In Cartesian coordinates with the z axis aligned with the beam, the magnetic-field component in the $x - y$ plane deflects the beam every λ_B interval in a different direction. We will include first the deflection due to the magnetic field component along the x axis then we will generalize to the $x - y$ plane. At the end we find that the angular distribution function is a Gaussian with azimuthal symmetry, hence the electrons and positrons distribution functions are equivalent, and it is sufficient to consider only one species.

At a given correlation-length interval denoted by i, the component of the magnetic field in the x direction ($B_{x,i} = B_{\rm IGM} \sin\theta' \cos\varphi'$) deflects the beam positrons along the y direction with a deflection angle

$$\Delta\theta_i(\theta', \varphi') = \frac{\lambda_B e B_{\rm IGM} \sin\theta' \cos\varphi'}{p}, \tag{A.1}$$

where p is the momentum of the beam particle and e is the elementary electric charge. $\Delta\theta_i$ is a random variable that depends on the random variables θ' and φ'. Since all the possible magnetic field orientations have the same probability, the mean deflection is

$$\begin{aligned}\mu &= \frac{1}{4\pi} \int_0^{\pi} \sin\theta' d\theta' \int_0^{2\pi} d\varphi' \Delta\theta_i(\theta', \varphi') P(\Delta\theta_i(\theta', \varphi')) \\ &= \frac{\lambda_B e B_{\rm IGM}}{4\pi p} \int_0^{\pi} d\theta' \int_0^{2\pi} d\varphi' \sin^2\theta' \cos\varphi' = 0,\end{aligned} \tag{A.2}$$

M. S. A. Alawashra, *Plasma Instabilities of TeV Pair Beams Induced by Blazars*, Springer Theses, https://doi.org/10.1007/978-3-031-98568-3

and the variance is

$$\sigma^2 = \frac{1}{4\pi}\left(\frac{\lambda_B e B_{\rm IGM}}{p}\right)^2 \int_0^{\pi} d\theta' \int_0^{2\pi} d\varphi' \sin^3\theta' \cos^2\varphi' \\ = \frac{1}{3}\left(\frac{\lambda_B e B_{\rm IGM}}{p}\right)^2. \tag{A.3}$$

The total deflection of the beam is computed as

$$\Delta\theta = \sum_{i=0}^{n} \Delta\theta_i, \tag{A.4}$$

where $n = \lambda_{\rm IC}/\lambda_B$ is the total number of the correlation lengths crossed by the beam during the its energy loss length (substituted here by the inverse Compton scattering length). Since n is very large in our case, $\lambda_{\rm IC} >> \lambda_B$, we can use the central limit theorem with $n \to \infty$ to find the distribution function of the total deflection,

$$f_{b,\theta_y}(\theta_y, p) = \frac{1}{\sqrt{2\pi}\, n\sigma} \exp -\frac{1}{2}\left(\frac{\theta_y}{\sqrt{n}\sigma}\right)^2, \quad -\pi \le \theta_y \le \pi, \tag{A.5}$$

which is a normal distribution with dispersion $\sqrt{n}\sigma$. For the magnetic-field component $B_y = B_{\rm IGM} \sin\theta' \cos\varphi'$, we get the same distribution for $f_{b,\theta_x}(\theta_x, p)$,

$$f_{b,\theta_x}(\theta_x, p) = \frac{1}{\sqrt{2\pi}\, n\sigma} \exp -\frac{1}{2}\left(\frac{\theta_x}{\sqrt{n}\sigma}\right)^2, \quad -\pi \le \theta_x \le \pi. \tag{A.6}$$

Note that by definition $n = \lambda_{\rm IC}/\lambda_{\rm B}$. Combining the two distributions in Eq. A.5 and Eq. A.6 using the result in Appendix A.2 we get the full angular distribution of the pair beam

$$f_{b,\theta}(\theta, p) = \frac{1}{\Delta\theta^2_{\rm IGMF}\pi} \exp -\left(\frac{\theta}{\Delta\theta_{\rm IGMF}}\right)^2; \; 0 \le \theta \le \pi, \; 0 \le \varphi \le 2\pi, \tag{A.7}$$

where

$$\Delta\theta_{\rm IGMF} = \frac{e B_{\rm IGMF}}{p}\sqrt{\frac{2}{3}\lambda_{\rm IC}\lambda_{\rm B}}. \tag{A.8}$$

What we have considered here is a fixed IGMF amplitude. Considering a IGMF with different amplitudes leads to the same result in terms of the root-mean-square IGMF with a numerical factor.

A.2 Transformation of $f_x(\theta_x) f_y(\theta_y)$ to $f(\theta, \varphi)$

Combining the distributions $f_x(\theta_x)$ and $f_y(\theta_y)$ to the distribution $f(\theta, \varphi)$ by

$$f(\theta, \varphi) \sin\theta d\theta d\varphi = f_x(\theta_x) f_y(\theta_y) d\theta_x d\theta_y, \tag{A.9}$$

where θ and φ are the spherical coordinate and θ_x and θ_y are defined in Fig. A.1. We rewrite Eq. A.9 using the Jacobian determinant

$$f(\theta, \varphi) = f_x(\theta_x(\theta, \varphi)) f_y(\theta_y(\theta, \varphi)) \frac{1}{\sin\theta} |\mathbf{J}(\theta, \varphi)|. \tag{A.10}$$

The relations between (θ_x, θ_y) and (θ, φ) can be found as follows; the Cartesian coordinates of a unit vector with (θ, φ) are $x = \sin\theta\cos\varphi$, $y = \sin\theta\sin\varphi$ and $z = \cos\theta$, then using the definitions of θ_x and θ_y in Fig. A.1

$$\tan\theta_x = \frac{x}{z} = \tan\theta\cos\varphi, \tag{A.11}$$

and

$$\tan\theta_y = \frac{y}{z} = \tan\theta\sin\varphi. \tag{A.12}$$

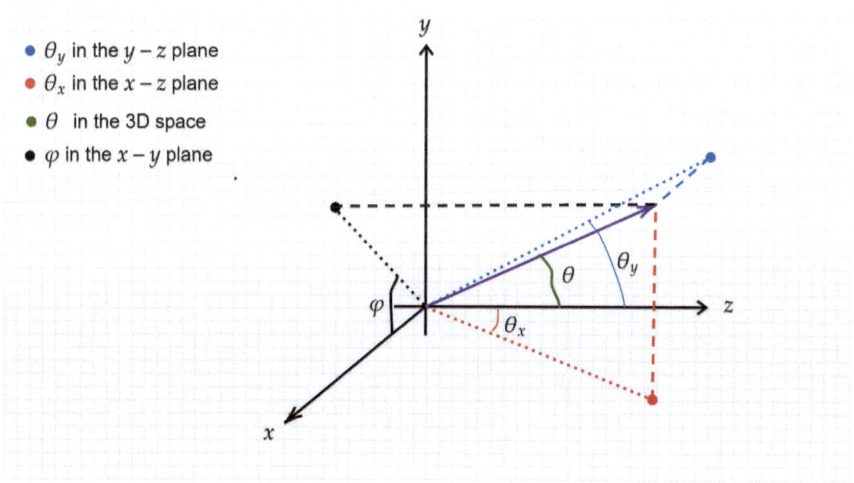

Fig. A.1 The two systems of coordinate for the angular discretion; the first one uses spherical coordinates, (θ, φ), and the second one is involves (θ_x, θ_y), where θ_y is the angle between the z axis and the projection on the $y - z$ plane and θ_x is the angle between the z axis and the projection on the $x - z$ plane

The Jacobian determinant is given by

$$
\begin{aligned}
|\mathbf{J}(\theta, \varphi)| &= \left| \frac{\partial \theta_x}{\partial \theta} \frac{\partial \theta_y}{\partial \varphi} - \frac{\partial \theta_x}{\partial \varphi} \frac{\partial \theta_y}{\partial \theta} \right| \\
&= \left| \frac{4 \tan \theta}{4 + \sin^2 \theta \tan^2 \theta \sin^2 2\varphi} \right| \approx |\tan \theta|,
\end{aligned}
\tag{A.13}
$$

for a small θ. Putting this in Eq. A.10 gives

$$
f(\theta, \varphi) = f_x(\theta_x(\theta, \varphi)) f_y(\theta_y(\theta, \varphi)) \frac{|\tan \theta|}{\sin \theta} \approx f_x(\theta_x(\theta, \varphi)) f_y(\theta_y(\theta, \varphi)), \tag{A.14}
$$

for a small θ.

Appendix B
Methodology of Fokker-Planck Simulation

B.1 Expansion for $\theta_{1,2}$

The resonance boundaries for the integration over the beam angle are given by

$$\cos\theta_{1,2} = \frac{\omega_p}{kv_b}\left(\cos\theta' \pm \sin\theta'\sqrt{\left(\frac{kv_b}{\omega_p}\right)^2 - 1}\right), \tag{B.1}$$

where $v_b \simeq c\left(1 - \frac{1}{2\gamma^2}\right)$. Dividing and multiplying the expression with kc/ω_p gives

$$\cos\theta_{1,2} = \left(\frac{\omega_p}{kc}\right)^2 \frac{1}{\left(1 - \frac{1}{2\gamma^2}\right)}\left(\frac{ck_{||}}{\omega_p} \pm \frac{ck_\perp}{\omega_p}\sqrt{\left(\frac{kv_b}{\omega_p}\right)^2 - 1}\right). \tag{B.2}$$

Expressing the parallel wave number as $k_{||} = \frac{\omega_p}{c}(1 + \varepsilon_{||})$ gives

$$\left(\frac{\omega_p}{kc}\right)^2 \frac{1}{\left(1 - \frac{1}{2\gamma^2}\right)} = \frac{1}{1 + \left(\frac{ck_\perp}{\omega_p}\right)^2 + 2\varepsilon_{||} + \varepsilon_{||}^2 - \frac{1}{2\gamma^2}\left(1 + \left(\frac{ck_\perp}{\omega_p}\right)^2 + 2\varepsilon_{||}\right)}, \tag{B.3}$$

and

$$\sqrt{\left(\frac{kv_b}{\omega_p}\right)^2 - 1} = \sqrt{\left(\frac{ck_\perp}{\omega_p}\right)^2 + 2\varepsilon_{||} + \varepsilon_{||}^2 - \frac{1}{\gamma^2}\left(1 + \left(\frac{ck_\perp}{\omega_p}\right)^2 + 2\varepsilon_{||}\right)}, \tag{B.4}$$

where we neglected the higher order terms of $\frac{\varepsilon_{||}^2}{2\gamma^2}$ that is always very small. After substituting Eqs. B.3 and B.4 in Eq. B.2, we get

M. S. A. Alawashra, *Plasma Instabilities of TeV Pair Beams Induced by Blazars*, Springer Theses, https://doi.org/10.1007/978-3-031-98568-3

$$\cos\theta_{1,2} = \left(\frac{1}{1+\left(\frac{ck_\perp}{\omega_p}\right)^2 + 2\varepsilon_{||} + \varepsilon_{||}^2 - \frac{1}{2\gamma^2}\left(1+\left(\frac{ck_\perp}{\omega_p}\right)^2 + 2\varepsilon_{||}\right)}\right)$$
$$\times\left(1+\varepsilon_{||} \pm \frac{ck_\perp}{\omega_p}\sqrt{\left(\frac{ck_\perp}{\omega_p}\right)^2 + 2\varepsilon_{||} + \varepsilon_{||}^2 - \frac{1}{\gamma^2}\left(1+\left(\frac{ck_\perp}{\omega_p}\right)^2 + 2\varepsilon_{||}\right)}\right). \tag{B.5}$$

We can see here in Eq. B.5, that the condition for the beam particles' speed to be relevant in the linear growth calculations is, $\left(\frac{ck_\perp}{\omega_p}\right)^2 + 2\varepsilon_{||} \leq \frac{10^3}{2\gamma^2}$. Equation B.5 can be rewritten as

$$\cos\theta_{1,2} = \left(\frac{1}{1+\left(\frac{ck_\perp}{\omega_p}\right)^2 + 2\varepsilon_{||} + \varepsilon_{||}^2 - \frac{1}{2\gamma^2}\left(1+\left(\frac{ck_\perp}{\omega_p}\right)^2 + 2\varepsilon_{||}\right)}\right)$$
$$\times\left(1+\varepsilon_{||} \pm \left(\frac{ck_\perp}{\omega_p}\right)^2\sqrt{1+\frac{2\varepsilon_{||}}{\left(\frac{ck_\perp}{\omega_p}\right)^2} + \frac{\varepsilon_{||}^2}{\left(\frac{ck_\perp}{\omega_p}\right)^2} - \frac{1}{\gamma^2\left(\frac{ck_\perp}{\omega_p}\right)^2}\left(1+\left(\frac{ck_\perp}{\omega_p}\right)^2 + 2\varepsilon_{||}\right)}\right). \tag{B.6}$$

Using the trigonometric identity, $\sin^2\frac{\theta}{2} = \frac{1-\cos\theta}{2}$, we then get

$$\sin^2\frac{\theta_{1,2}}{2} = \left(\frac{1}{2\left(1+\left(\frac{ck_\perp}{\omega_p}\right)^2 + 2\varepsilon_{||} + \varepsilon_{||}^2 - \frac{1}{2\gamma^2}\left(1+\left(\frac{ck_\perp}{\omega_p}\right)^2 + 2\varepsilon_{||}\right)\right)}\right)$$
$$\times\left(\varepsilon_{||} + \varepsilon_{||}^2 - \frac{1}{2\gamma^2}\left(1+\left(\frac{ck_\perp}{\omega_p}\right)^2 + 2\varepsilon_{||}\right)\right.$$
$$\left. + \left(\frac{ck_\perp}{\omega_p}\right)^2\left[1 \mp \sqrt{1+\frac{2\varepsilon_{||}}{\left(\frac{ck_\perp}{\omega_p}\right)^2} + \frac{\varepsilon_{||}^2}{\left(\frac{ck_\perp}{\omega_p}\right)^2} - \frac{1}{\gamma^2\left(\frac{ck_\perp}{\omega_p}\right)^2}\left(1+\left(\frac{ck_\perp}{\omega_p}\right)^2 + 2\varepsilon_{||}\right)}\right]\right). \tag{B.7}$$

We see in Eqs. B.6 and B.7 that for $\left(\frac{ck_\perp}{\omega_p}\right)^2 \gg 2\varepsilon_{||}$ and $\gamma^2\left(\frac{ck_\perp}{\omega_p}\right)^2 \gg 1$, the square root becomes very close to 1 making the numerical round-off of the square root affecting the calculations. We solved this problem by expanding the square root up to the following orders

$$\sqrt{1+\frac{2\varepsilon_{||}}{\left(\frac{ck_\perp}{\omega_p}\right)^2}+\frac{\varepsilon_{||}^2}{\left(\frac{ck_\perp}{\omega_p}\right)^2}-\frac{1}{\gamma^2\left(\frac{ck_\perp}{\omega_p}\right)^2}\left(1+\left(\frac{ck_\perp}{\omega_p}\right)^2+2\varepsilon_{||}\right)}\approx 1+\frac{\varepsilon_{||}}{\left(\frac{ck_\perp}{\omega_p}\right)^2}+\frac{1}{2}\frac{\varepsilon_{||}^2}{\left(\frac{ck_\perp}{\omega_p}\right)^2}$$
$$-\frac{1}{2}\frac{\varepsilon_{||}^2}{\left(\frac{ck_\perp}{\omega_p}\right)^4}-\frac{1}{2}\frac{\varepsilon_{||}^3}{\left(\frac{ck_\perp}{\omega_p}\right)^4}+\frac{1}{2}\frac{\varepsilon_{||}^3}{\left(\frac{ck_\perp}{\omega_p}\right)^6}-\frac{1}{2\gamma^2\left(\frac{ck_\perp}{\omega_p}\right)^2}\left(1+\left(\frac{ck_\perp}{\omega_p}\right)^2+2\varepsilon_{||}\right) \tag{B.8}$$
$$-\frac{1}{8\gamma^4\left(\frac{ck_\perp}{\omega_p}\right)^4}\left(1+\left(\frac{ck_\perp}{\omega_p}\right)^4+2\left(\frac{ck_\perp}{\omega_p}\right)^2\right)+\frac{\varepsilon_{||}}{2\gamma^2\left(\frac{ck_\perp}{\omega_p}\right)^4}\left(1+2\left(\frac{ck_\perp}{\omega_p}\right)^2\right),$$

for $\left(\frac{ck_\perp}{\omega_p}\right)^2 \geq 200\varepsilon_{||}$ and $\gamma^2\left(\frac{ck_\perp}{\omega_p}\right)^2 \geq 100$.

Substituting the result of Eqs. B.8 in B.7, gives

$$\sin^2\frac{\theta_{1,2}}{2}=\left(\frac{1}{2\left(1+\left(\frac{ck_\perp}{\omega_p}\right)^2+2\varepsilon_{||}+\varepsilon_{||}^2-\frac{1}{2\gamma^2}\left(1+\left(\frac{ck_\perp}{\omega_p}\right)^2+2\varepsilon_{||}\right)\right)}\right)$$
$$\times\left(\left(\frac{ck_\perp}{\omega_p}\right)^2+\varepsilon_{||}+\varepsilon_{||}^2-\frac{1}{2\gamma^2}\left(1+\left(\frac{ck_\perp}{\omega_p}\right)^2+2\varepsilon_{||}\right)\right.$$
$$\mp\left[\left(\frac{ck_\perp}{\omega_p}\right)^2+\varepsilon_{||}+\frac{\varepsilon_{||}^2}{2}\left(1-\frac{1}{\left(\frac{ck_\perp}{\omega_p}\right)^2}\right)\right.$$
$$+\frac{\varepsilon_{||}^3}{2}\left(-\frac{1}{\left(\frac{ck_\perp}{\omega_p}\right)^2}+\frac{1}{\left(\frac{ck_\perp}{\omega_p}\right)^4}\right)-\frac{1}{2\gamma^2}\left(1+\left(\frac{ck_\perp}{\omega_p}\right)^2+2\varepsilon_{||}\right)$$
$$\left.\left.-\frac{1}{8\gamma^4\left(\frac{ck_\perp}{\omega_p}\right)^2}\left(1+\left(\frac{ck_\perp}{\omega_p}\right)^4+2\left(\frac{ck_\perp}{\omega_p}\right)^2\right)+\frac{\varepsilon_{||}}{2\gamma^2\left(\frac{ck_\perp}{\omega_p}\right)^2}\left(1+2\left(\frac{ck_\perp}{\omega_p}\right)^2\right)\right]\right), \tag{B.9}$$

yielding the following for the resonance lower angle bound θ_1

$$\sin\frac{\theta_1}{2}=\frac{\frac{\varepsilon_{||}}{2\left(\frac{ck_\perp}{\omega_p}\right)}}{\left(1+\left(\frac{ck_\perp}{\omega_p}\right)^2+2\varepsilon_{||}+\varepsilon_{||}^2-\frac{1}{2\gamma^2}\left(1+\left(\frac{ck_\perp}{\omega_p}\right)^2+2\varepsilon_{||}\right)\right)^{1/2}}$$
$$\times\left(1+\left(\frac{ck_\perp}{\omega_p}\right)^2+\varepsilon_{||}\left(1-\frac{1}{\left(\frac{ck_\perp}{\omega_p}\right)^2}\right)\right.$$
$$\left.+\frac{1}{4\gamma^4\varepsilon_{||}^2}\left(1+\left(\frac{ck_\perp}{\omega_p}\right)^4+2\left(\frac{ck_\perp}{\omega_p}\right)^2\right)-\frac{1}{\gamma^2\varepsilon_{||}}\left(1+2\left(\frac{ck_\perp}{\omega_p}\right)^2\right)\right)^{1/2}, \tag{B.10}$$

and the following for the resonance upper angle bound θ_2

$$\sin\frac{\theta_2}{2} = \frac{1}{\left(1+\left(\frac{ck_\perp}{\omega_p}\right)^2 + 2\varepsilon_{||} + \varepsilon_{||}^2 - \frac{1}{2\gamma^2}\left(1+\left(\frac{ck_\perp}{\omega_p}\right)^2 + 2\varepsilon_{||}\right)\right)^{1/2}}$$
$$\times\left(\left(\frac{ck_\perp}{\omega_p}\right)^2 + \varepsilon_{||} - \frac{1}{2\gamma^2}\left(1+\left(\frac{ck_\perp}{\omega_p}\right)^2 + 2\varepsilon_{||}\right)\right.$$
$$+\frac{\varepsilon_{||}^2}{4}\left(3-\frac{1}{\left(\frac{ck_\perp}{\omega_p}\right)^2}\right) + \frac{\varepsilon_{||}^3}{4}\left(-\frac{1}{\left(\frac{ck_\perp}{\omega_p}\right)^2} + \frac{1}{\left(\frac{ck_\perp}{\omega_p}\right)^4}\right)$$
$$\left. - \frac{1}{16\gamma^4\left(\frac{ck_\perp}{\omega_p}\right)^2}\left(1+\left(\frac{ck_\perp}{\omega_p}\right)^4 + 2\left(\frac{ck_\perp}{\omega_p}\right)^2\right) + \frac{\varepsilon_{||}}{4\gamma^2\left(\frac{ck_\perp}{\omega_p}\right)^2}\left(1+2\left(\frac{ck_\perp}{\omega_p}\right)^2\right)\right)^{1/2}. \tag{B.11}$$

We used Eqs. B.10 and B.11 to calculate the values of $\theta_{1,2}$, when $\left(\frac{ck_\perp}{\omega_p}\right)^2 \geq 200\varepsilon_{||}$ and $\gamma^2\left(\frac{ck_\perp}{\omega_p}\right)^2 \geq 100$, and Eq. B.7 otherwise.

B.2 Parallel Modes Transformation

The goal of this section is to transform the linear growth rate (Eq. 5.2) using the transformation of Eq. 5.10 for the case of the parallel modes, $\left(\frac{ck_\perp}{\omega_p}\right)^2 \leq \left(\frac{ck_{||}}{\omega_p} - 1\right)$. Using the values of $\cos\theta_{1,2}$ given by Eq. B.5

$$\cos\theta_{1,2} = \left(\frac{1}{1+\left(\frac{ck_\perp}{\omega_p}\right)^2 + 2\varepsilon_{||} + \varepsilon_{||}^2 - \frac{1}{2\gamma^2}\left(1+\left(\frac{ck_\perp}{\omega_p}\right)^2 + 2\varepsilon_{||}\right)}\right)$$
$$\times\left(1+\varepsilon_{||} \pm \frac{ck_\perp}{\omega_p}\sqrt{\left(\frac{ck_\perp}{\omega_p}\right)^2 + 2\varepsilon_{||} + \varepsilon_{||}^2 - \frac{1}{\gamma^2}\left(1+\left(\frac{ck_\perp}{\omega_p}\right)^2 + 2\varepsilon_{||}\right)}\right), \tag{B.12}$$

and the approximation $\cos\theta \approx 1 - \frac{\theta^2}{2}$ that is always valid for the blazar-induced TeV pairs, we get

$$\cos\theta_1 - \cos\theta = \left(\frac{1}{1 + \left(\frac{ck_\perp}{\omega_p}\right)^2 + 2\varepsilon_{||} + \varepsilon_{||}^2 - \frac{1}{2\gamma^2}\left(1 + \left(\frac{ck_\perp}{\omega_p}\right)^2 + 2\varepsilon_{||}\right)} \right) \times \left[\left(\frac{\theta^2}{2} + \frac{1}{2\gamma^2}\right)\left(1 + \left(\frac{ck_\perp}{\omega_p}\right)^2 + 2\varepsilon_{||}\right) - \varepsilon_{||} - \left(\frac{ck_\perp}{\omega_p}\right)^2 - \varepsilon_{||}^2 + \frac{ck_\perp}{\omega_p}\sqrt{\left(\frac{ck_\perp}{\omega_p}\right)^2 + 2\varepsilon_{||} + \varepsilon_{||}^2 - \frac{1}{\gamma^2}\left(1 + \left(\frac{ck_\perp}{\omega_p}\right)^2 + 2\varepsilon_{||}\right)} \right], \tag{B.13}$$

and

$$\cos\theta - \cos\theta_2 = \left(\frac{1}{1 + \left(\frac{ck_\perp}{\omega_p}\right)^2 + 2\varepsilon_{||} + \varepsilon_{||}^2 - \frac{1}{2\gamma^2}\left(1 + \left(\frac{ck_\perp}{\omega_p}\right)^2 + 2\varepsilon_{||}\right)} \right) \times \left[-\left(\frac{\theta^2}{2} + \frac{1}{2\gamma^2}\right)\left(1 + \left(\frac{ck_\perp}{\omega_p}\right)^2 + 2\varepsilon_{||}\right) + \varepsilon_{||} + \left(\frac{ck_\perp}{\omega_p}\right)^2 + \varepsilon_{||}^2 + \frac{ck_\perp}{\omega_p}\sqrt{\left(\frac{ck_\perp}{\omega_p}\right)^2 + 2\varepsilon_{||} + \varepsilon_{||}^2 - \frac{1}{\gamma^2}\left(1 + \left(\frac{ck_\perp}{\omega_p}\right)^2 + 2\varepsilon_{||}\right)} \right]. \tag{B.14}$$

Where we neglected the quantities $\frac{\varepsilon_{||}^2}{2\gamma^2}$, $\theta^2\varepsilon_{||}^2$ and $\varepsilon_{||}^3$ that are always negligible. Multiplying Eqs. B.13 and B.14 we get the following

$$(\cos\theta_1 - \cos\theta)(\cos\theta - \cos\theta_2) = \frac{1 + \left(\frac{ck_\perp}{\omega_p}\right)^2 + 2\varepsilon_{||}}{\left(1 + \left(\frac{ck_\perp}{\omega_p}\right)^2 + 2\varepsilon_{||} + \varepsilon_{||}^2 - \frac{1}{2\gamma^2}\left(1 + \left(\frac{ck_\perp}{\omega_p}\right)^2 + 2\varepsilon_{||}\right)\right)^2} \times \left[-\left(\frac{\theta^2}{2} + \frac{1}{2\gamma^2}\right)^2\left(1 + \left(\frac{ck_\perp}{\omega_p}\right)^2 + 2\varepsilon_{||}\right) + \theta^2\left(\varepsilon_{||} + \left(\frac{ck_\perp}{\omega_p}\right)^2\right) - \varepsilon_{||}^2 + \frac{\varepsilon_{||}}{\gamma^2} \right]. \tag{B.15}$$

In the regime of $\varepsilon_{||} \sim \left(\frac{ck_\perp}{\omega_p}\right)^2$ and $\varepsilon_{||} >> \left(\frac{ck_\perp}{\omega_p}\right)^2$, one can see that $\frac{ck_\perp}{\omega_p} << 1$ since always $\varepsilon_{||} << 1$. We also have $\frac{1}{2\gamma^2} << 1$ for the blazar-induced TeV beams. Using those limits Eq. B.15 simplify to

$$(\cos\theta_1 - \cos\theta)(\cos\theta - \cos\theta_2) = -\frac{\theta^4}{4} + \theta^2\left(\left(\frac{ck_\perp}{\omega_p}\right)^2 + \varepsilon_{||} - \frac{1}{2\gamma^2}\right) - \left(\varepsilon_{||} - \frac{1}{2\gamma^2}\right)^2. \tag{B.16}$$

Using the transformation of the variable, θ, to the following variable, λ,

$$\lambda = \frac{\theta^2}{2\left(\left(\frac{ck_\perp}{\omega_p}\right)^2 + \varepsilon_{||} - \frac{1}{2\gamma^2}\right)}, \tag{B.17}$$

Equation B.16 simplify to

$$(\cos\theta_1 - \cos\theta)(\cos\theta - \cos\theta_2) = \left(\left(\frac{ck_\perp}{\omega_p}\right)^2 + \varepsilon_{||} - \frac{1}{2\gamma^2}\right)^2 (\lambda - \lambda_1)(\lambda_2 - \lambda), \tag{B.18}$$

where

$$\lambda_{1,2} = 1 \mp \frac{ck_\perp/\omega_p}{\left(\left(\frac{ck_\perp}{\omega_p}\right)^2 + \varepsilon_{||} - \frac{1}{2\gamma^2}\right)} \sqrt{\left(\frac{ck_\perp}{\omega_p}\right)^2 + 2\varepsilon_{||} - \frac{1}{\gamma^2}}. \tag{B.19}$$

Using again the transformation B.17, the numerator of Eq. 5.2 transform as

$$\left(-2f_b(p,\theta)\sin\theta + (\cos\theta - \frac{kv_b}{\omega_p}\cos\theta')\frac{\partial f_b(p,\theta)}{\partial\theta}\right)d\theta =$$
$$\left(\left(\frac{ck_\perp}{\omega_p}\right)^2 + \varepsilon_{||} - \frac{1}{2\gamma^2}\right)\left(-2f_b(p,\lambda) + \left(-\lambda + \frac{\frac{1}{2\gamma^2} - \varepsilon_{||}}{\left(\left(\frac{ck_\perp}{\omega_p}\right)^2 + \varepsilon_{||} - \frac{1}{2\gamma^2}\right)}\right)\frac{\partial f_b(p,\lambda)}{\partial\lambda}\right)d\lambda. \tag{B.20}$$

Finally combining the results in Eqs. B.18 and B.20, 5.2 transforms to

$$I(p) = \int_{\lambda_1}^{\lambda_2} d\lambda \frac{F_\lambda(p,\lambda)}{[(\lambda - \lambda_1)(\lambda_2 - \lambda)]^{1/2}}, \tag{B.21}$$

where

$$F_\lambda(p,\lambda) = -2f_b(p,\lambda) + \left(-\lambda + \frac{\frac{1}{2\gamma^2} - \varepsilon_{||}}{\left(\left(\frac{ck_\perp}{\omega_p}\right)^2 + \varepsilon_{||} - \frac{1}{2\gamma^2}\right)}\right)\frac{\partial f_b(p,\lambda)}{\partial\lambda}, \tag{B.22}$$

and $\lambda_{1,2}$ are given by Eq. B.19.

B.3 Diffusion Coefficients

The diffusion coefficients are given by

$$D_{ij} = \pi e^2 \int d^3\mathbf{k} W(\mathbf{k}) \frac{k_i k_j}{k^2} \delta(\mathbf{k} \cdot \mathbf{v}_b - \omega_p), \tag{B.23}$$

where the unstable wave wavevector $\mathbf{k} = (k, \theta', \varphi')$ and the beam particles velocity $\mathbf{v}_b = (v_b \simeq c(1 - \frac{1}{2\gamma^2}), \theta, \varphi = 0)$ are both defined in the spherical coordinates with the beam propagation axis being the z-axis. Because of the azimuth symmetry, we set $\varphi = 0$ without losing the generality yielding

$$\begin{aligned} D_{ij} =& \pi e^2 \int k^2 dk \int d\cos\theta' \int d\varphi' W(\mathbf{k}, t) \frac{k_i k_j}{k^2} \\ & \times \delta(kc(1 - \frac{1}{2\gamma^2})[\sin\theta' \sin\theta \cos\varphi' + \cos\theta' \cos\theta] \quad \omega_p). \end{aligned} \tag{B.24}$$

After transforming the delta function we get

$$D_{ij} = \pi e^2 \int k^2 dk \int d\cos\theta' \int d\varphi' W(\mathbf{k}, t) \frac{k_i k_j}{k^2} \frac{\delta(\varphi' - \varphi'_*)}{kc(1 - \frac{1}{2\gamma^2}) \sin\theta' \sin\theta \sin\varphi'_*}, \tag{B.25}$$

where $\cos\varphi'_* = \frac{\omega_p/(kc(1-\frac{1}{2\gamma^2})) - \cos\theta' \cos\theta}{\sin\theta' \sin\theta}$.

k_i is the projection of wave-vector ($\mathbf{k} = k \sin\theta' \cos\varphi' \hat{\mathbf{x}} + k \sin\theta' \sin\varphi' \hat{\mathbf{y}} + k \cos\theta' \hat{\mathbf{z}}$) to the spatial direction i. We have fixed the azimuth angle of the pair beam to zero ($\varphi = 0$), therefore we have only the beam modulus momentum (p) and the angler direction $\hat{\theta} = \cos\theta \hat{\mathbf{x}} - \sin\theta \hat{\mathbf{z}}$. Based on this we find that k_p is the modulus of the wave-vector and $k_\theta = \mathbf{k} \cdot \hat{\theta} = k[\sin\theta' \cos\theta \cos\varphi' - \cos\theta' \sin\theta]$.

Substituting the values of k_p and k_θ and integrating over φ', gives

$$\begin{aligned} \begin{Bmatrix} D_{pp} \\ D_{p\theta} \\ D_{\theta\theta} \end{Bmatrix} =& \pi \frac{m_e \omega_p^2}{n_e} \int_{\omega_p/c}^{\infty} k^2 dk \int_{\cos\theta'_1}^{\cos\theta'_2} d\cos\theta' \\ & \times \frac{W(\mathbf{k})}{kc(1 - \frac{1}{2\gamma^2}) \sqrt{(\cos\theta' - \cos\theta'_1)(\cos\theta'_2 - \cos\theta')}} \begin{Bmatrix} 1 \\ \xi \\ \xi^2 \end{Bmatrix}, \end{aligned} \tag{B.26}$$

where

$$\xi = \sin\theta' \cos\varphi'_* \cos\theta - \cos\theta' \sin\theta = \frac{\cos\theta \frac{\omega_p}{kc(1-\frac{1}{2\gamma^2})} - \cos\theta'}{\sin\theta}. \tag{B.27}$$

and the boundaries of $\cos\theta'$ are fixed by the condition

$$\left|\cos\varphi'_*\right| = \left|\frac{\omega_p/(kc(1-\frac{1}{2\gamma^2})) - \cos\theta'\cos\theta}{\sin\theta'\sin\theta}\right| \leq 1, \tag{B.28}$$

which gives

$$\cos\theta'_{1,2} = \frac{\omega_p}{kc(1-\frac{1}{2\gamma^2})}\left[\cos\theta \pm \sin\theta\sqrt{\left(\frac{kc}{\omega_p}\right)^2(1-\frac{1}{2\gamma^2})^2 - 1}\right]. \tag{B.29}$$

Since we have the calculations for the linear growth rate in the Cartesian coordinates $(k_\perp, \varepsilon_{||})$ where $k_{||} = \frac{\omega_p}{c}(1+\varepsilon_{||})$, we need to transform the diffusion coefficients integrand from the coordinates $(k, \cos\theta')$ to the coordinates $(k_\perp, \varepsilon_{||})$. Using the Jacobian determinant $|J| = \frac{\omega_p}{c}\frac{k_\perp}{k^2}$ in Eq. B.26, we get

$$\begin{Bmatrix} D_{pp} \\ D_{p\theta} \\ D_{\theta\theta} \end{Bmatrix} = \pi\frac{m_e\omega_p^2}{n_e c}\int d\varepsilon_{||}\int_{k_{\perp,1}}^{\infty} dk_\perp k_\perp \times \frac{W(k_\perp, \varepsilon_{||})}{\sqrt{\theta^2\left(\frac{k_\perp}{\omega_p/c}\right)^2 + \varepsilon_{||}\left[\theta^2 + \frac{1}{\gamma^2}\right] - \varepsilon_{||}^2 - \left[\frac{1}{2\gamma^2} + \frac{\theta^2}{2}\right]^2}}\begin{Bmatrix} 1 \\ \xi \\ \xi^2 \end{Bmatrix}, \tag{B.30}$$

where for $\theta << 1$ and $\varepsilon_{||} << 1$, we can approximate ξ as

$$\xi = -\frac{\omega_p}{kc}\frac{1}{\theta}\left[\frac{\theta^2}{2} + \varepsilon_{||} - \frac{1}{2\gamma^2}\right], \tag{B.31}$$

and we used the relation

$$(kc/\omega_p)\sqrt{(\cos\theta' - \cos\theta'_1)(\cos\theta'_2 - \cos\theta')} = \sqrt{\theta^2\left(\frac{k_\perp}{\omega_p/c}\right)^2 + \varepsilon_{||}\left[\theta^2 + \frac{1}{\gamma^2}\right] - \varepsilon_{||}^2 - \left[\frac{1}{2\gamma^2} + \frac{\theta^2}{2}\right]^2}. \tag{B.32}$$

The resonance boundaries (Eq. B.29) translated to a lower bound on $k_\perp$ for a given $\varepsilon_{||}$. The modes with negative $\varepsilon_{||}$ are stable and therefore we are only left with the lower limit for the positive $\varepsilon_{||}$ that is given by

$$k_{\perp,1} = \frac{\omega_p}{c}\frac{1}{\theta}\sqrt{\varepsilon_{||}^2 + \frac{1}{4\gamma^4} + \frac{\theta^4}{4} - \frac{\varepsilon_{||}}{\gamma^2} + \frac{1}{2}\left(\frac{\theta}{\gamma}\right)^2 - \varepsilon_{||}\theta^2}. \tag{B.33}$$

In order to have a proper numerical girding over the unstable waves spectrum we transform the coordinates from $(k_\perp, \varepsilon_{||})$ to $(k_\perp, \theta^R)$ where $\theta^R = \frac{\varepsilon_{||}}{ck_\perp/\omega_p}$ finding the following final expression for the diffusion coefficients

$$\begin{Bmatrix} D_{pp} \\ D_{p\theta} \\ D_{\theta\theta} \end{Bmatrix} = \pi \frac{m_e \omega_p^2}{n_e c\theta} \int_{R(\theta,\gamma)} dk_\perp k_\perp \int_{R(\theta,\gamma)} d\theta^R \times \frac{W(k_\perp, \theta^R)}{\sqrt{1 - \left(\frac{\theta^R}{\theta}\right)^2 + \frac{\theta^R}{ck_\perp/\omega_p}\left[1 + \left(\frac{1}{\gamma\theta}\right)^2\right] - \left(\frac{\omega_p}{ck_\perp}\right)^2 \left[\frac{1}{2\gamma^2\theta} + \frac{\theta}{2}\right]^2}} \begin{Bmatrix} 1 \\ \xi \\ \xi^2 \end{Bmatrix}, \tag{B.34}$$

where

$$\xi = -\frac{1}{\sqrt{1 + 2\theta^R (ck_\perp/\omega_p) + (ck_\perp/\omega_p)^2 (1 + \theta^{R^2})}} \left[\frac{\theta^R}{\theta}\frac{ck_\perp}{\omega_p} + \frac{\theta}{2} - \frac{1}{2\theta\gamma^2}\right], \tag{B.35}$$

and the resonance region $R(\theta, \gamma)$ is defined by the following condition

$$\left(\frac{ck_\perp}{\omega_p}\right)^2 \left(\theta^2 - \theta^{R^2}\right) + \frac{ck_\perp}{\omega_p}\theta^R \left[\theta^2 + \frac{1}{\gamma^2}\right] - \left[\frac{1}{2\gamma^2} + \frac{\theta^2}{2}\right]^2 \geq 0. \tag{B.36}$$

Appendix C
Von Neumann Stability Analysis of the Diffusion Equation

C.1 FTCS Numerical Scheme

The Fokker–Planck diffusion equation we solve in Sect. 5.1.3 reads

$$\frac{\partial f(p,\theta)}{\partial t} = \frac{1}{p^2\theta^2}\frac{\partial}{\partial \ln\theta}\left(D_{\theta\theta}\frac{\partial}{\partial \ln\theta} f(p,\theta)\right). \tag{C.1}$$

Applying the explicit forward time-centered space scheme (FTCS), we get

$$\frac{f_i^{t+1} - f_i^t}{\Delta t} = \frac{1}{p^2\theta_i^2(\Delta \ln\theta)^2}\left[D^t_{\theta\theta i+\frac{1}{2}}(f^t_{i+1} - f^t_i) - D^t_{\theta\theta i-\frac{1}{2}}(f^t_i - f^t_{i-1})\right], \tag{C.2}$$

where we defined $f_i^t = f(p,\theta_i,t)$ and $D^{t+1}_{\theta\theta i+\frac{1}{2}} \equiv D_{\theta\theta}(\theta_{i+\frac{1}{2}}, t+\Delta t)$ with t being the initial time, Δt is the time step and $\theta_{i+\frac{1}{2}} = \exp{(\ln\theta)_{i+\frac{1}{2}}}$. We compute $(\ln\theta)_{i+\frac{1}{2}}$ as $(\ln\theta)_{i+\frac{1}{2}} = ((\ln\theta)_i + (\ln\theta)_{i+1})/2$ with $\Delta \ln\theta$ as the grid spacing in logarithmic of θ.

Neglecting the numerical errors in the diffusion coefficients and defining the round-off error in the beam distribution as

$$\varepsilon_i^t = f_i^t - E_i^t, \tag{C.3}$$

where E_i^t is the exact solution in the absence of the round-off error and f_i^t is the numerical solution. Since both E_i^t and f_i^t satisfy the Eq. C.2 then the error also does

$$\frac{\varepsilon_i^{t+1} - \varepsilon_i^t}{\Delta t} = \frac{1}{p^2\theta_i^2(\Delta \ln\theta)^2}\left[D^t_{\theta\theta i+\frac{1}{2}}(\varepsilon^t_{i+1} - \varepsilon^t_i) - D^t_{\theta\theta i-\frac{1}{2}}(\varepsilon^t_i - \varepsilon^t_{i-1})\right]. \tag{C.4}$$

M. S. A. Alawashra, *Plasma Instabilities of TeV Pair Beams Induced by Blazars*, Springer Theses, https://doi.org/10.1007/978-3-031-98568-3

Writing the error as a Fourier sum

$$\varepsilon_i^t \equiv \varepsilon(\theta_i, t) = \sum_{k_m} E_{k_m}(t) e^{ik_m\theta_i}, \tag{C.5}$$

where the time dependence of the error is included in the amplitude of the error Fourier components $E_{k_m}(t)$. As the Eq. C.4 is linear, we can pull the sum out and the equation is satisfied for every Fourier component k_m. To find the error evolution with time, we substitute the Fourier modes of the error for every grid point

$$\begin{aligned}
\varepsilon_{i,k_m}^t &= E_{k_m}(t) e^{ik_m\theta_i}, \\
\varepsilon_{i,k_m}^{t+1} &= E_{k_m}(t + \Delta t) e^{ik_m\theta_i}, \\
\varepsilon_{i-1,k_m}^t &= E_{k_m}(t) e^{ik_m(\theta_i - \Delta\theta_i)}, \\
\varepsilon_{i+1,k_m}^t &= E_{k_m}(t) e^{ik_m(\theta_i + \Delta\theta_i)},
\end{aligned} \tag{C.6}$$

in Eq. C.4, yielding the amplification factor

$$\begin{aligned}
G = \frac{E_{k_m}(t+\Delta t)}{E_{k_m}(t)} = 1 + \alpha(\theta_i)\Big[&(D^t_{\theta\theta i+\frac{1}{2}} + D^t_{\theta\theta i-\frac{1}{2}})(\cos(k_m\Delta\theta_i) - 1) \\
&+ i(D^t_{\theta\theta i+\frac{1}{2}} - D^t_{\theta\theta i-\frac{1}{2}})\sin(k_m\Delta\theta_i)\Big],
\end{aligned} \tag{C.7}$$

where

$$\alpha(\theta_i) = \frac{\Delta t}{p^2\theta_i^2(\Delta\ln\theta)^2}. \tag{C.8}$$

Multiplying with the complex conjugate of the amplification factor, G, we get

$$\begin{aligned}
|G|^2 = &\Big[1 + \alpha(\theta_i)(D^t_{\theta\theta i+\frac{1}{2}} + D^t_{\theta\theta i-\frac{1}{2}})(\cos(k_m\Delta\theta_i) - 1)\Big]^2 \\
&+ \alpha^2(\theta_i)(D^t_{\theta\theta i+\frac{1}{2}} - D^t_{\theta\theta i-\frac{1}{2}})^2\sin^2(k_m\Delta\theta_i).
\end{aligned} \tag{C.9}$$

The condition for the numerical stability is that the amplification factor is smaller than one, $|G|^2 \leq 1$. If this equality is satisfied when $\sin(k_m\Delta\theta_i) = 1$ then it's satisfied globally, and so the numerical stability condition reduces to

$$\frac{(D'^2_{\theta\theta i+\frac{1}{2}} + D'^2_{\theta\theta i-\frac{1}{2}})}{(D^t_{\theta\theta i+\frac{1}{2}} + D^t_{\theta\theta i-\frac{1}{2}})}\alpha(\theta_i) \leq 1, \tag{C.10}$$

yielding the following condition for numerically stable time step

$$\Delta t \leq \min_i \left[p^2\theta_i^2(\Delta\ln\theta)^2 \frac{(D^t_{\theta\theta i+\frac{1}{2}} + D^t_{\theta\theta i-\frac{1}{2}})}{(D'^2_{\theta\theta i+\frac{1}{2}} + D'^2_{\theta\theta i-\frac{1}{2}})}\right]. \tag{C.11}$$

C.2 Crank–Nicolson Numerical Scheme

The semi-implicit CrankâŁ"Nicolson scheme of the diffusion equation reads as

$$\frac{f_i^{t+1}-f_i^t}{\Delta t} = \frac{1}{2p^2\theta_i^2(\Delta\ln\theta)^2}\Big[D^t_{\theta\theta i+\frac{1}{2}}(f^t_{i+1}-f^t_i) - D^t_{\theta\theta i-\frac{1}{2}}(f^t_i - f^t_{i-1}) + D^{t+1}_{\theta\theta i+\frac{1}{2}}(f^{t+1}_{i+1}-f^{t+1}_i) - D^{t+1}_{\theta\theta i-\frac{1}{2}}(f^{t+1}_i - f^{t+1}_{i-1})\Big]. \tag{C.12}$$

This case is much more complicated to analyse in its current form, we can simplify the analysis by considering the case, $D^{t+1}_{\theta\theta i+\frac{1}{2}} \approx D^t_{\theta\theta i+\frac{1}{2}}$. Applying the Fourier analysis of the round-off error as in the previous section we get the following relation for the error amplification factor, G,

$$L \equiv \frac{G-1}{G+1} = \beta(\theta_i)\left[\frac{i}{4\theta_i}\sin(k_m\Delta\theta_i)\left(1+\frac{\partial\ln D^t_{\theta\theta i}}{\partial\ln\theta_i}\right) - \frac{2}{\theta_i\Delta\ln\theta}\sin^2\left(\frac{k_m\Delta\theta_i}{2}\right)\right], \tag{C.13}$$

where

$$\beta(\theta_i) = \frac{2\Delta t D^t_{\theta\theta i}}{p^2\theta_i\Delta\ln\theta}. \tag{C.14}$$

Writing $L = \mathrm{Re}(L) + i\mathrm{Im}(L)$, where $\mathrm{Re}(L)$ is the real part of L and $\mathrm{Im}(L(\theta_i))$ is the imaginary part of L, we can write the amplification factor as

$$G = \frac{1+L}{1-L} = \frac{1+\mathrm{Re}(L)+i\mathrm{Im}(L)}{1-\mathrm{Re}(L)-i\mathrm{Im}(L)}. \tag{C.15}$$

Multiplying and dividing the amplification factor with the complex conjugate of its denominator and then multiplying the result with the complex conjugate of the amplification factor we get

$$|G|^2 = \frac{\left[1-\mathrm{Re}^2(L)-\mathrm{Im}^2(L)\right]^2 + 4\mathrm{Im}^2(L)}{\left[(1-\mathrm{Re}(L))^2+\mathrm{Im}^2(L)\right]^2}. \tag{C.16}$$

Applying the numerical condition $|G|^2 \le 1$, we get the following

$$\frac{7}{4}\mathrm{Re}(L) - \mathrm{Re}^2(L) - \mathrm{Im}^2(L) \le 1. \tag{C.17}$$

As $\mathrm{Re}(L)$ is always negative, then we get always something negative on the left-hand side of the equality and therefore this scheme is always numerically stable. Therefore, we don't have any constrain on the choice of Δt and $\Delta\ln\theta$ using the Crank-Nicolson scheme.

Appendix D
Beam Parallel Momentum Distribution

We approximated the pair beam parallel momentum distribution function found in [1] with a logarithmic Gaussian at Lorentz factors higher than 6×10^6. This replaces the step-function cut-off with an exponential one. This additional function has continuity in derivative and value at the transition point with the distribution found in [1], where the resulting distribution function is given by

$$
\begin{aligned}
f_\gamma(\gamma) =& N_1 \left(\frac{\gamma}{\gamma_1}\right)^{-b1} \exp -\sqrt{\frac{\gamma_1}{\gamma}} \Theta\left[(\gamma - 6 \times 10^3)(6 \times 10^6 - \gamma)\right] \\
&+ N_2 \left(\frac{\gamma}{\gamma_2}\right)^{-\frac{\ln(\gamma/\gamma_2)}{b_2} - 1} \Theta\left[(\gamma - 6 \times 10^6)(10^8 - \gamma)\right],
\end{aligned}
\tag{D.1}
$$

where the parameters are summarized in Table D.1. We have plotted the pair beam distribution function in Fig. D.1.

Table D.1 The parameters for the approximation in Eq. D.1

i	b_i	γ_i	N_i
1	1.60	1.58×10^6	3.00×10^{-7}
2	5.78	1.55×10^6	1.14×10^{-7}

M. S. A. Alawashra, *Plasma Instabilities of TeV Pair Beams Induced by Blazars*, Springer Theses, https://doi.org/10.1007/978-3-031-98568-3

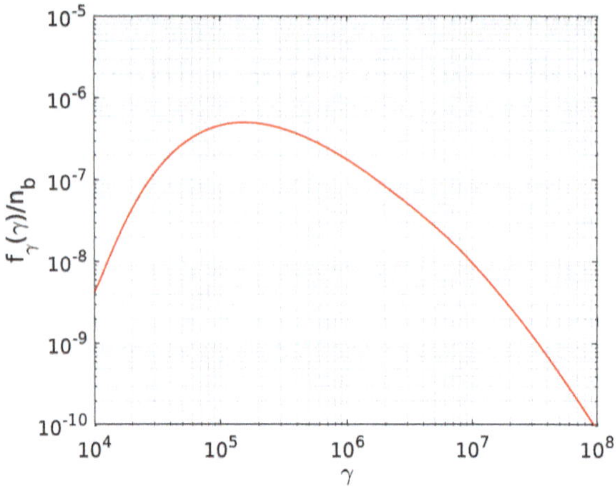

Fig. D.1 The parallel momentum distribution of the pair beam that we have used in the study of Chap. 5 as it is given by Eq. D.1

Reference

1. S. Vafin, I. Rafighi, M. Pohl, and J. Niemiec. The electrostatic instability for realistic pair distributions in blazar/ebl cascades. *The Astrophysical Journal*, 857(1):43, Apr 2018. ISSN 1538-4357. https://doi.org/10.3847/1538-4357/aab552. URL http://dx.doi.org/10.3847/1538-4357/aab552

MIX
Papier aus verantwortungsvollen Quellen
Paper from responsible sources
FSC® C105338

If you have any concerns about our products,
you can contact us on
ProductSafety@springernature.com

In case Publisher is established outside the EU,
the EU authorized representative is:
Springer Nature Customer Service Center GmbH
Europaplatz 3, 69115 Heidelberg, Germany

Printed by Libri Plureos GmbH
in Hamburg, Germany